Roberto Conti (Ed.)

Calculus of Variations, Classical and Modern

Lectures given at a Summer School of the Centro Internazionale Matematico Estivo (C.I.M.E.), held in Bressanone (Bolzano), Italy, June 10-18, 1966

Springer

FONDAZIONE CIME ROBERTO CONTI

C.I.M.E. Foundation
c/o Dipartimento di Matematica "U. Dini"
Viale margagni n. 67/a
50134 Firenze
Italy
cime@math.unifi.it

ISBN 978-3-642-11041-2 ISBN 978-3-642-11042-9 (eBook)
DOI 10.1007/978-3-642-11042-9
Springer Heidelberg Dordrecht London New York

Reprint of the 1st ed. C.I.M.E., Ed. Cremonese, Roma 1967
With kind permission of C.I.M.E.

Printed on acid-free paper

Springer.com

CENTRO INTERNAZIONALE MATEMATICO ESTIVO
(C.I.M.E.)
1° - Ciclo - Bressanone dal 10 al 18 giugno
1966

"CALCULUS OF VARIATIONS, CLASSICAL AND MODERN"

Coordinatore : Prof. R. CONTI

A. BLAQUIERE : Quelques apsects geometriques des processus optimaux. pag. 1

Ch. CASTAING : Quelques problèmes de mesurabilité liés a la théorie des commandes. pag. 75

L. CESARI : "Existence theorems for Lagrange and Pontryagin problems of the calculus of variations and optimal control more dimensional extensions in Sobolev space". pag. 85

H. HALKIN : Optimal control as programmin in infinite dimensional spaces pag. 177

C. OLECH : "The range of integrals of a certain class vector-valued functions" pag. 195

E. H. ROTHE : Weak topology and calculus of variations pag. 205

E. O. ROXIN : Problems about the set of attainability " . 239

CENTRO INTERNAZIONALE MATEMATICO ESTIVO
(C.I.M.E.)

Austin BLAQUIERE

QUELQUES ASPECTS GEOMETRIQUES DES PROCESSUS OPTIMAUX

Corso tenuto a Bressanone dal 10 al 18 giugno 1966

QUELQUES ASPECTS GEOMETRIQUES DES PROCESSUS OPTIMAUX[1)]

par

Austin BLAQUIERE [2)]

I

DEFINITION DES SURFACES LIMITES, PROPRIETES GLOBALES DE CES SURFACES.

1. Introduction

Nous considèrerons un système dynamique dont l'état, au temps t, est représenté par le point $\mathfrak{x} = (x_1, x_2, \dots x_n)$ dans un espace Euclidien à n-dimensions, E^n. Nous supposerons que le comportement du système, c'est-à-dire l'évolution au cours du temps de ses variables d'état, dépend du choix de certaines règles dans un ensemble règles donné. Pour le moment il ne sera pas nécessaire de préciser les propriétés de cet ensemble de règles; par contre, nous ferons quelques hypothèses en ce qui concerne le comportement du système.

A toute règle admissible, c'est-à-dire appartenant à l'ensemble donné, correspond une évolution au cours du temps de l'état du système, $\mathfrak{x}(t)$. En d'autres termes, étant donnée la règle $\mathfrak{r}$, le point représentatif de l'état se déplace le long d'un trajet p dans

---

1) Cet exposé prend pour base deux rapports : University of California IER Report AM-64-10 , et AM-65-11, On the Geometry of Optimal Processes -- Part I et Part II, par A. Blaquière et G. Leitmann.

2) Professeur a la Faculté des Sciences de Paris, Laboratoire d'Automatique Théorique.

l'espace des états.

Dans la discussion suivante nous considèrerons les trajets qui sont associés a toutes les règles admissibles.
En particulier nous fixerons notre attention sur les règles qui transfèrent le système d'un état donné, $\underset{\sim}{x}$, à une variété terminale, θ^1, de dimension inferieure à n. En général, certaines règles transfèrent le système de $\underset{\sim}{x}'$ à $\underset{\sim}{x}^1 \in \theta^1$, tandis que les autres correspondent à des trajets qui n'atteignent pas la variété terminale. Ces derniers ont pour point terminal $\bar{\underset{\sim}{x}}^1 \notin \theta^1$.

Nous nous donnerons d'autre part un index de performance, c'est-à-dire une fonctionnelle, qui fait correspondre un nombre réel unique à chaque transfert associé à une règle admissible. Ce nombre sera appelé coût du transfert. En général le coût d'un transfert de $\underset{\sim}{x}$ à $\underset{\sim}{x}^f$, où $\underset{\sim}{x}^f = \underset{\sim}{x}^1$ ou $\bar{\underset{\sim}{x}}^1$, dépend de la règle r et du trajet p correspondant; nous le désignerons par $V(\underset{\sim}{x}, \underset{\sim}{x}^f; r, p)$.

Nous dirons qu'une règle r^* est optimale si le trajet correspondant, p^*, issu de $\underset{\sim}{x}$, a pour point terminal $\underset{\sim}{x}^{1*} \in \theta^1$ et si le coût est minimum, c'est-à-dire si

$$(1) \qquad V(\underset{\sim}{x}, \underset{\sim}{x}^{1*}; r^*, p^*) \leq V(\underset{\sim}{x}, \underset{\sim}{x}^1; r, p)$$

pour toutes les règles admissibles.

Notons qu'une règle optimale n'est pas nécessairement unique, par contre le coût minimum est unique; nous poserons donc

$$V^*(\boldsymbol{x} \,;\, \theta^1) \triangleq V(\boldsymbol{x}, \boldsymbol{x}^{1*} \,;\, r^*, p^*) \qquad \forall r^*$$

ce qui met en évidence le fait que, pour une variété terminale donnée θ^1, le coût minimum est fonction seulement de l'état initial.

Considérons maintenant deux sous-ensembles de l'espace des états E^n :

(i) E, ensemble des points tels que de chacun d'eux est issu un trajet qui atteint θ^1 ; c'est-à-dire

$$E \triangleq \left\{ \boldsymbol{x} : \exists\, p \text{ de } \boldsymbol{x} \text{ à } \theta^1 \right\}$$

(ii) E^* , ensemble des points tels que de chacun d'eux est issu un trajet associé à une règle optimale; c'est-à-dire

$$E^* \triangleq \left\{ \boldsymbol{x} : \exists\, p^* \text{ de } \boldsymbol{x} \text{ à } \theta^1 \right\}$$

Nous introduirons les hypothèses suivantes :

l'hypothèse d'additivité des coûts,[1] c'est - à -dire

$$V(\boldsymbol{x}, \boldsymbol{x}^f ; r, p) = V(\boldsymbol{x}, \boldsymbol{x}^i ; r, p^i) + V(\boldsymbol{x}^i, \boldsymbol{x}^f \; r, p_i)$$

(2) $$\forall\, \boldsymbol{x}^i \in p \;, \qquad p = p^i \cup p_i$$

et

$$\lim_{\boldsymbol{x}^i \to \boldsymbol{x}^f} V(\boldsymbol{x}^i, \boldsymbol{x}^f ; r, p_i) = 0$$

---

[1] Cette hypothèse sera valable pour les coûts intégraux considérés dans la suite.

2. Trajectoires dans l'espace des états augmenté

Considérons maintenant un espace des états augmenté, l'espace Euclidien à (n+1)-dimensions, E^{n+1}, dont les points seront désignés par $\underset{\sim}{x} = (x_0, \underset{\approx}{x})$; et dans cet espace une courbe Γ, dont la projection dans E^n est le trajet p.

Par définition

(3) $$\Gamma = \left\{ \underset{\sim}{x}^i : x_0^i + V(\underset{\approx}{x}^i, \underset{\approx}{x}^f ; r, p_i) = C , p_i \subset p \right\}$$

où p est le trajet associé à la règle r, et C un paramètre constant . Γ sera appelée trajectoire dans l'espace des états augmenté. Une trajectoire Γ^* associée à une règle optimale sera appelée trajectoire optimale

(4) $$\Gamma^* \triangleq \left\{ \underset{\sim}{x}^i : x_0^i + V(\underset{\approx}{x}^i, \underset{\approx}{x}^{1*} ; r^*, p_i) = C , p_i \subset p^* \right\}$$

En faisant varier le paramètre C dans (3) et (4), on engendre une famille de trajectoires à 1 paramètre. Les trajectoires de la famille appartiennent à une surface cylindrique, de génératrices parallèles à l'axe x_o, dont l'intersection avec E^n est p ou p^* .

3. Surface limites et surfaces isocoût optimales

$V^*(\underset{\approx}{x} ; \theta^1)$ est défini sur E^*, de telle sorte que l'équation

(5) $$x_0 + V^*(\underset{\approx}{x} ; \theta^1) = C$$

définit une surface Σ dans E^{n+1}. Comme $V^*(\underset{\sim}{x}\,;\,\theta^1)$ est une fonction univoque de $\underset{\sim}{x}$, Σ est une surface à une seule nappe. En d'autres termes, il existe une correspondance 1-1 entre les points de Σ et les points de E^* : toute parallèle à l'axe x_o qui rencontre E^* perce Σ en un point et un seul.

En général, la fonction $x_o = C - V^*(\underset{\sim}{x}\,;\,\theta^1)$ s'annule sur une surface S dont l'équation est [1]

$$V^*(\underset{\sim}{x}\,;\,\theta^1) = C \tag{6}$$

En faisant varier le paramètre C dans (5) et (6), on engendre deux familles de surfaces, chacune à 1 paramètre : les surfaces limites $\{\Sigma\}$ dans $E^* \times x_o$, et les surfaces isocoût optimales $\{S\}$ dans E^*. La première appellation sera clairement expliquée par l'étude des propriétés globales de ces surfaces, au paragraphe suivant. La seconde provient du fait que chaque surface S est le lieu des points initiaux correspondant au <u>même</u> coût de transfert minimum C, pour une variété θ^1 donnée.

4. Quelques propriétés des surfaces limites

De la propriété d'additivité des coûts (2) et de la définition de l'optimalité (1), il résulte que

---

1) Dans le cas exceptionnel où $V^*(\underset{\sim}{x}\,;\,\theta^1)$ est indépendant de $\underset{\sim}{x}$, on peut montrer que $V^*(\underset{\sim}{x}\,;\theta^1) \equiv 0$. Dans ce cas la surface S n'est pas définie.

$$(7) \qquad V(\underset{\sim}{x}^i, \underset{\sim}{x}^{1*}; r^*, p_i) = V^*(\underset{\sim}{x}^i; \theta^1) \; \forall \underset{\sim}{x}^i \in p^*, \quad p_i \subset p^*$$

Par conséquent

Lemme 1. Toute trajectoire optimale , Γ^*, qui rencontre la variété $\theta^1 \times x_0$ au point $\underset{\sim}{x}^1$ est entièrement contenue dans la surface Σ passant par ce point.

De la définition (5) il résulte que les surfaces de la famille $\{\Sigma\}$ se déduisent les unes des autres par une translation parallèle à l'axe x_o , et sont ordonnées suivant cet axe de la même facon que les valeurs de C . Ainsi, par tout point de $E^* \times x_o$ passe une surface Σ et une seule. Cette propriété et le Lemme 1 conduisent au Lemme 2 :

Lemme 2. Toute trajectoire optimale, Γ^*, dont un point appartient à une surface limite Σ , est entièrement contenue dans Σ ; ainsi, les surfaces limites sont les lieux des trajectoires optimales.

Une surface limite Σ donnée sépare $E^* \times x_o$ en deux régions ouvertes , relativement à Σ . Nous désignerons ces régions par A/Σ ("above" Σ) et B/Σ ("below" Σ) , respectivement. Si Σ est la surface limite correspondant au coût C :

$$A/\Sigma \triangleq \{ \underset{\sim}{x} \; : \; x_o > C - V^*(\underset{\sim}{x} \; ; \theta^1) \}$$

et

$$B/\Sigma \triangleq \{ \underset{\sim}{x} \; : \; x_o < C - V^*(\underset{\sim}{x} \; ; \theta^1 \;) \}$$

Si $\underset{\sim}{x} \in A/\Sigma$, nous dirons qu'il est un A-point par rapport à Σ ; de même ,
si $\underset{\sim}{x} \in B/\Sigma$, nous dirons qu'il est un B-point par rapport à Σ .

Le Lemme 2 et les définitions des trajectoires et des surfaces Σ nous conduisent au Lemme 3 .

Lemme 3. Il n'existe pas de trajectoire issue d'un point appartenant à une surface limite donnée, Σ et qui rencontre $\theta^1 \times x_o$ en un B-point par rapport à Σ.

Enfin des Lemmes 2 et 3 , nous déduisons le Théorème fondamental suivant

Théorème I. Une trajectoire (optimale ou non-optimale) dont le point initial appartient à une surface limite donnée, Σ ne peut avoir un B-point par rapport à Σ.

Ce théorème explique pourquoi nous avons appelé de telles surfaces; surfaces limites; nous voyons en effet qu'une surface Σ donnée appartient à la frontière de la région qui contient toutes les trajectoires issues des points de Σ .

Remarquons que la propriété d'additivité des coûts est la seule hypothèse sur laquelle s'appuie le Théorème I .

5. Equations d'état

Dans la suite nous limiterons la discussion au cas de systèmes dont les variables d'état sont des fonctions du temps qui vérifient les équations d'état

$$\frac{dx_j}{dt} = f_j\,(x_1, x_2, \ldots x_n, u_1, u_2, \ldots u_m)\,, \quad j = 1, 2, \ldots n \tag{8}$$

où $u_1, u_2, \ldots u_m$ sont les variables de commande qui définissent un point $\underset{\sim}{u}$ dans un espace Euclidien à m-dimensions, E^m. Nous dirons que la commande (ou le vecteur de commande) $\underset{\sim}{u} = \underset{\sim}{u}(t)$, $t_0 \leqslant t \leqslant t_1$, est admissible, si

(i) $\underset{\sim}{u}(t)$ est défini et continu par morceaux sur l'intervalle $[t_0, t_1]$;

(ii) $\underset{\sim}{u}(t) \in U$ quel que soit $t \in [t_0, t_1]$, où U est un sous-ensemble constant [1)] de E^m.

Nous supposerons que les fonctions $f_j(\underset{\sim}{x}, \underset{\sim}{u})$ et $\partial f_j(\underset{\sim}{x}, \underset{\sim}{u}) / \partial x_i$, i, j = 1, 2, ... n, sont définies et continues sur $E^n \times U$. Ainsi, lorsque la commande admissible $\underset{\sim}{u}(t)$ est donnée, $t_0 \leqslant t \leqslant t_1$, ainsi que la condition initiale $\underset{\sim}{x} = \underset{\sim}{x}^{\circ}$ au temps $t = t_0$, il existe une solution unique, continue, $\underset{\sim}{x}(t)$. Les équations d'état, et les commandes admissibles, définissent l'ensemble des règles admissibles.

6. Critère de coût intégral

Dans la suite nous considèrerons un index de performance (ou critère de coût) integral, de la forme

$$\int_{t_0}^{t_1} f_0(\underset{\sim}{x}(t), \underset{\sim}{u}(t))\, dt \tag{9}$$

On a dans ce cas, avec les notations du paragraphe 1

$$V(\underset{\sim}{x}^i, \underset{\sim}{x}^f; r, p_i) = \int_t^{t_1} f_0(\underset{\sim}{x}(\tau), \underset{\sim}{u}(\tau))\, d\tau$$

1) Tous les résultats s'étendent sans difficulté au cas des systèmes pour lesquels U dépend de l'état.

et

$$x_o(t) + \int_t^{t_1} f_o(\mathfrak{x}(\tau) ; \underset{\sim}{u}(\tau))\, d\tau = C$$

Par conséquent

$$(10) \qquad \frac{dx_o}{dt} = f_o(\mathfrak{x}, \underset{\sim}{u})$$

Les fonctions $f_o(\mathfrak{x}, \underset{\sim}{u})$ et $\partial f_o(\mathfrak{x}, \underset{\sim}{u}) / \partial x_j$, $j = 1, 2, \ldots n$, sont supposées définies et continues sur $E^n \times U$.

En combinant (8) et (10) nous avons l'équation dont les solutions sont les trajectoires dans E^{n+1} , soit

$$(11) \qquad \frac{d\underset{\sim}{x}}{dt} = \underset{\sim}{f}(\underset{\sim}{x}, \underset{\sim}{u})$$

où

$$\underset{\sim}{f}(\underset{\sim}{x}; \underset{\sim}{u}) = (f_o(\mathfrak{x}, \underset{\sim}{u}), f_1(\mathfrak{x}, \underset{\sim}{u}), \ldots f_n(\mathfrak{x}, \underset{\sim}{u}))$$

Une <u>commande optimale</u> $\underset{\sim}{u}^*(t)$, $t_o \leq t \leq t_1$, assure le transfert du système d'un état donné $\underset{\sim}{\mathfrak{x}}^o$, au temps t_o , à un état $\underset{\sim}{\mathfrak{x}}^1 \in \theta^1$, au temps t_1 , et ce transfert est tel que l'intégrale (9) ait sa valeur minimale.

Notons que le temps de transfert $t_1 - t_o$ n'est pas spécifié ici . Tous les résultats s'étendent sans difficulté au cas des problèmes non-autonomes.

<u>7. Propriétés d'unc transformation linéaire</u>

Considérons un vecteur $\underset{\sim}{\eta} = (\eta_o, \eta_1, \ldots \eta_n)$ au point $\underset{\sim}{x}^*(t)$ de la trajectoire optimale Γ^* pilotée par la commande $\underset{\sim}{u}^*(t)$, $t_o \leq t \leq t_1$; et supposons que ses composantes

$\eta_j = \eta_j(t)$ soient solutions des <u>équations variationnelles</u>

(12) $$\frac{d\eta_j}{dt} = \sum_{i=0}^{n} \left. \frac{\partial f_j(\underset{\sim}{x}, \underset{\sim}{u}^*(t))}{\partial x_i} \right|_{\underset{\sim}{x} = \underset{\sim}{x}^*(t)} \eta_i \quad , \quad j = 0, 1, \ldots n$$

pour des conditions initiales données $\eta_j(t_o) = \eta_j^o$.

Les équations (12) définissent une transformation linéaire, non-singulière, $A(t_o, t)$, telle que

(13) $$\underset{\sim}{\eta}(t) = A(t_o, t)\underset{\sim}{\eta}^o, \quad t_o \leqslant t \leqslant t_1$$

Des propriétés d'une telle transformation, il résulte que

<u>Lemme 4.</u> Le transformé, $\Pi(\underset{\sim}{x}^*(t))$, d'un plan $\Pi(\underset{\sim}{x}^*(t_o))$ contenant le point $\underset{\sim}{x}^*(t_o)$ de Γ^*, a les propriétés suivantes :

(i) $\Pi(\underset{\sim}{x}^*(t))$ est défini quel que soit t, $t_o \leqslant t \leqslant t_1$

(ii) $\Pi(\underset{\sim}{x}^*(t))$ est un plan de même dimension que $\Pi(\underset{\sim}{x}^*(t_o))$;

(iii) la direction de la normale au plan $\Pi(\underset{\sim}{x}^*(t))$ varie de façon continue sur $[t_o, t_1]$.

8. Transformation du plan tangent

Un point d'une surface sera dit <u>fortement régulier</u> si le plan tangent à la surface en ce point est défini. Désignons par Γ^* une trajectoire optimale, de commande $\underset{\sim}{u}^*(t)$ $t_o \leqslant t \leqslant t_1$, dont le point initial, $\underset{\sim}{x}^*(t_o)$, est un point interieur[1], fortement

1) C'est-à-dire un point interieur de $E^* \times x_o$

régulier, de la surface limite $\sum$. D'après le Lemme 2, Γ^* est entièrement contenue dans $\sum$.

Considérons <u>dans $\sum$</u> un voisinage du point $\underset{\sim}{x}^*(t_o)$ défini par

(14) $$\Delta(\underset{\sim}{x}^*(t_o)) \triangleq \left\{ \underset{\sim}{x} \; ; \; \underset{\sim}{x} = \underset{\sim}{x}(t_o) \in \sum \right\}$$

où

$$\underset{\sim}{x}(t_o) = \underset{\sim}{x}^*(t_o) + \varepsilon \underset{\sim}{\eta}^\circ + \underset{\sim}{0}(\varepsilon)$$

et

(i) ε est assez petit pour que $\underset{\sim}{x}(t_o) \in \sum$;

(ii) $$\lim_{\varepsilon \to 0} \frac{\underset{\sim}{0}(\varepsilon)}{\varepsilon} = 0$$

(iii) $\underset{\sim}{\eta}^\circ \in T_\Delta(\underset{\sim}{x}^*(t_o))$, le plan tangent à $\Delta(\underset{\sim}{x}^*(t_o))$ en $\underset{\sim}{x}^*(t_o)$.

Il est clair que $T_\Delta(\underset{\sim}{x}^*(t_o)) \equiv T_\Sigma(\underset{\sim}{x}^*(t_o))$, où $T_\Sigma(\underset{\sim}{x}^*(t_o))$ est le plan tangent à $\sum$ en $\underset{\sim}{x}^*(t_o)$.

La transformation de $\Delta(\underset{\sim}{x}^*(t_o))$, au moyen des équations des trajectoires (11) avec $\underset{\sim}{u} = \underset{\sim}{u}^*(t)$, $t_o \leqslant t \leqslant t_1$, conduit à l'ensemble de points

(15) $$\Delta(\underset{\sim}{x}^*(t)) \triangleq \left\{ \underset{\sim}{x} \; : \; \underset{\sim}{x} = \underset{\sim}{x}(t) \right\}, \quad t_o \leqslant t \leqslant t_1$$

où

$$\underset{\sim}{x}(t) = \underset{\sim}{x}^*(t) + \varepsilon \underset{\sim}{\eta}(t) + \underset{\sim}{0}(t, \varepsilon)$$

et

(i) $\dfrac{\underset{\sim}{0}(t,\varepsilon)}{\varepsilon} \to 0$ uniformement quel que soit $t \in [t_o, t_1]$ quand $\varepsilon \to 0$;

(ii) $\underset{\sim}{\eta}(t)$ est une solution de (12) avec $\underset{\sim}{\eta}(t_o) = \underset{\sim}{\eta}^\circ$.

Comme chaque point $\underset{\sim}{x}(t)$ appartient à une trajectoire dont le point initial est dans Σ , il résulte du Théorème I que $\underset{\sim}{x}(t) \in A/\Sigma \cup \Sigma$, pourvu que $\underset{\sim}{x}^*(t)$ soit un point interieur de Σ et que $|\varepsilon|$ soit suffisamment petit.

Ce résultat et le Lemme 4 conduisent au Lemme 5.

Lemme 5. Si les points $\underset{\sim}{x}^*(t_o)$ et $\underset{\sim}{x}^*(t)$, $t_o \leqslant t \leqslant t_1$, de la trajectoire optimale Γ^* sur la surface limite Σ sont des points interieurs, fortement réguliers, de Σ , le plan tangent à Σ en $\underset{\sim}{x}^*(t_o)$ admet pour transformé le plan tangent à Σ en $\underset{\sim}{x}^*(t)$, dans la transformation linéaire définie par les équations variationnelles (12); c'est-à-dire

(16) $$T_\Sigma\ (\underset{\sim}{x}^*(t)\ ) = A\ (t_o,\ t)\ T_\Sigma\ (\underset{\sim}{x}^*(t_o)\ )$$

9. Points interieurs fortement réguliers d'une surface limite

Désignons par $\underset{\sim}{n}(\underset{\sim}{x}')$ le vecteur unité qui est normal à la surface limite Σ , et dirigé vers la région B/Σ , en un point interieur fortement régulier $\underset{\sim}{x}'$ de Σ ; c'est-à-dire

(17) $$n_o(\underset{\sim}{x}') \leqslant 0$$

où n_o est la composante de $\underset{\sim}{n}$ suivant l'axe x_o. Comme conséquence directe du Théorème I, nous avons

(18) $$\underset{\sim}{n}\ (\underset{\sim}{x}')\ .\ \underset{\sim}{f}\ (\underset{\sim}{x}'\ ,\ \underset{\sim}{u}\ ) \leqslant\ 0 \qquad \forall \underset{\sim}{u} \in U$$

De plus, si $\underset{\sim}{x}' = \underset{\sim}{x}^*(t')$ appartient à la trajectoire optimale Γ^* associée à la commande $\underset{\sim}{u}^*(t)$, $t_o \leqslant t \leqslant t_1$, on a d'après le Lemme 2

(19) $\underset{\sim}{n}(\underset{\sim}{x}^*(t'))\,.\,\underset{\sim}{f}(\underset{\sim}{x}^*(t'),\ \underset{\sim}{u}^*(t')) = 0$

10. Trajectoires optimales fortement régulières

Supposons que Γ^* -- correspondant à $\underset{\sim}{x}^*(t)$ et $\underset{\sim}{u}^*(t)$, $t_o \leqslant t \leqslant t_1$ -- soit une trajectoire optimale fortement régulière, c'est-à-dire une trajectoire dont tous les points soient points interieurs, fortement réguliers, de la surface limite Σ, à laquelle appartient Γ^*.

Considérons maintenant les équations adjointes des équations variationnelles (12), soit

(20) $$\frac{d\lambda_j}{dt} = -\sum_{i=0}^{n} \left.\frac{\partial f_i(\underset{\sim}{x},\ \underset{\sim}{u}^*(t))}{\partial x_j}\right|_{\underset{\sim}{x}=\underset{\sim}{x}^*(t)} \lambda_i\ , \quad j = 0, 1, \ldots n$$

La solution $\underset{\sim}{\lambda}(t) = (\lambda_o(t), \lambda_1(t), \ldots \lambda_n(t))$ de ces équations adjointes, pour des conditions initiales données $\underset{\sim}{\lambda}(t_o) = \underset{\sim}{\lambda}^o$, est unique et continue sur $[t_o, t_1]$.

Nous avons le résultat bien connu

(21) $\underset{\sim}{\lambda}(t)\,.\,\eta(t) = \text{constante} \quad \forall\, t \in [t_o, t_1]$

Choisissons les conditions initiales suivantes pour (12) et (20), respectivement :

(i) $\eta^o \neq 0$ et $\in T_\Sigma(\underset{\sim}{x}^*(t_o))$

(ii) $\underset{\sim}{\lambda}^o \neq 0$, et $\perp T_\Sigma(\underset{\sim}{x}^*(t_o))$ et dirigé vers B/Σ.

On a alors d'après (21)

(22) $\underset{\sim}{\lambda}(t)\,.\,\eta(t) = 0 \quad \forall\, t \in [t_o, t_1]$

d'où , en utilisant (18) et (19)

(23) $\underset{\sim}{\lambda}(t) \cdot \underset{\sim}{f}(\underset{\sim}{x}^*(t), \underset{\sim}{u}) \leqslant 0 \qquad \forall \underset{\sim}{u} \in U$

et

(24) $\underset{\sim}{\lambda}(t) \cdot \underset{\sim}{f}(\underset{\sim}{x}^*(t), \underset{\sim}{u}^*(t)) = 0$

quel que soit t sur $[t_o, t_1]$

Comme, x_o n'intervient pas explicitement dans (20), et compte tenu de (17), on a

(25) $\lambda_o(t) \equiv$ constante $\leqslant 0$

11. Condition de transversalité terminale

Rappelons que nous nous proposons de transfèrer le système, d'équations d'état (8) et d'index de performance (9), d'un état donné, $\underset{\sim}{x}^\circ$ au temps t_o, à une variété terminale donnée, θ^1 au temps non spécifié t_1. Nous supposerons que θ^1 est définie par l'intersection de q surfaces, $1 \leqslant$ $\leqslant q \leqslant n$, d'équations

(26) $\theta^1_\sigma(\underset{\sim}{x}) = 0 \quad , \quad \sigma = 1, 2, \ldots q$

De plus, nous supposerons que grad $\theta^1_\sigma(\underset{\sim}{x})$, $\sigma = 1, 2, \ldots q$ est défini pour tout $\underset{\sim}{x} \in \theta^1$, et que ces vecteurs grad $\theta^1_\sigma(\underset{\sim}{x})$ sont linéairement indépendants pour tout $\underset{\sim}{x} \in \theta^1$. Ainsi θ^1 possède un plan tangent unique $T_{\theta^1}(\underset{\sim}{x})$ en chacun de ses points. Notons d'ailleurs que θ^1 est une variété de dimension n-q

Soit Σ la surface limite correspondant au coût C, et soit $\Theta \triangleq \theta^1 \times x_o$. Alors le plan tangent $T_{\Sigma \cap \Theta^1}(x^1)$ à l'intersection $\Sigma \cap \Theta^1$, en $\underset{\sim}{x}^1 = (C, \underset{\sim}{\mathfrak{x}}^1)$, est l'ensemble des points $\underset{\sim}{x} = (C, \underset{\sim}{\mathfrak{x}})$, où $\underset{\sim}{\mathfrak{x}} \in T_{\theta^1}(\underset{\sim}{\mathfrak{x}}^1)$.

Il est bien clair que

$$T_{\Sigma \cap \Theta^1}(\underset{\sim}{x}^1) \subset T_{\Sigma}(\underset{\sim}{x}^1)$$

(Fig. 1)

Donc, d'après(22), $\underset{\sim}{\lambda}(t_1)$ est normal à $T_{\Sigma \cap \Theta^1}(x^*(t_1))$ c'est-à-dire normal à <u>tout</u> vecteur $\eta^1 = (\eta_o^1, \eta_1^1, \ldots \eta_n^1) \in T_{\Sigma \cap \Theta^1}(x^*(t_1))$.

Ainsi nous aboutissons à la <u>condition de transversalité terminale</u>

$$(27) \qquad \underset{\sim}{\lambda}(t_1) \cdot \eta^1 = 0$$

où

$$\eta_o^1 = 0$$

$$(28) \qquad \sum_{j=1}^{n} \frac{\partial \theta_\sigma^1(\underset{\sim}{\mathfrak{x}})}{\partial x_j} \Bigg|_{\underset{\sim}{x} = \underset{\sim}{\mathfrak{x}}^{1*}} \eta_j = 0, \ \sigma = 1, 2, \ldots q$$

Compte tenu de l'indépendance linéaire des vecteurs grad $\theta_\sigma^1(\underset{\sim}{\mathfrak{x}})$, on peut résoudre (28) et déterminer q+1 composantes de η^1 comme combinaisons linéaires de n-q composantes arbitraires.

Dans (27), les coefficients des composantes arbitraires de η^1 doivent s'annuler, ce qui fournit n-q conditions sur $\underset{\sim}{\lambda}(t)$ et $\underset{\sim}{\mathfrak{x}}(t)$ au point terminal d'une trajectoire optimale Γ^*.

Notons que $\lambda_o(t_1)$ n'apparait pas dans ces conditions. Ainsi, la condition de transversalité terminale, jointe aux conditions terminales (26), fournit un ensemble de n relations entre 2n variables. n autres relations sont fournies par la spécification de l'état initial.

12. Condition de transversalité initiale

Jusqu'ici nous avons supposé que l'état initial , $\underset{\sim}{\mathfrak{x}}^{o}$ est prescrit. Si on astreint simplement l'état initial à appartenir à une variété donnée, θ^0 , moyennant des conditions sur θ^o analogues à celles que nous avons imposées plus haut sur θ^1, on aboutit à une condition de transversalité initiale analogue à celle que nous avons obtenue dans le précédent paragraphe.

Supposons que θ^o soit l'intersection de p surfaces, $1 \leqslant p \leqslant n$, d'équations

$$(29) \qquad \theta^o_\rho(\underset{\sim}{\mathfrak{x}}) = 0 \ , \qquad \rho = 1, 2, \ldots p$$

où , par hypothèse, les vecteurs grad $\theta^o_\rho(\underset{\sim}{\mathfrak{x}})$, $\rho = 1, 2, \ldots p$ existent et sont linéairement indépendants, de telle sorte que θ^o a un plan tangent unique , $T_{\theta^o}(\underset{\sim}{\mathfrak{x}})$, en $\underset{\sim}{\mathfrak{x}} \in \theta^o$.

Considérons maintenant deux points, $\underset{\sim}{\mathfrak{x}}^o_1$ et $\underset{\sim}{\mathfrak{x}}^o_2$, appartenant tous deux à E^* et situés sur θ^o. Désignons par Γ^*_1 et Γ^*_2 des trajectoires optimales issues de $\underset{\sim}{x}^o_1 = (0, \underset{\sim}{\mathfrak{x}}^o_1)$ et $\underset{\sim}{x}^o_2 = (0, \underset{\sim}{\mathfrak{x}}^o_2)$, respectivement.

Soient Σ_1 et Σ_2 les surfaces limites correspondantes, de paramètres C_1 et C_2, respectivement. Supposons que $\underset{\sim}{x}^o_1$ soit un

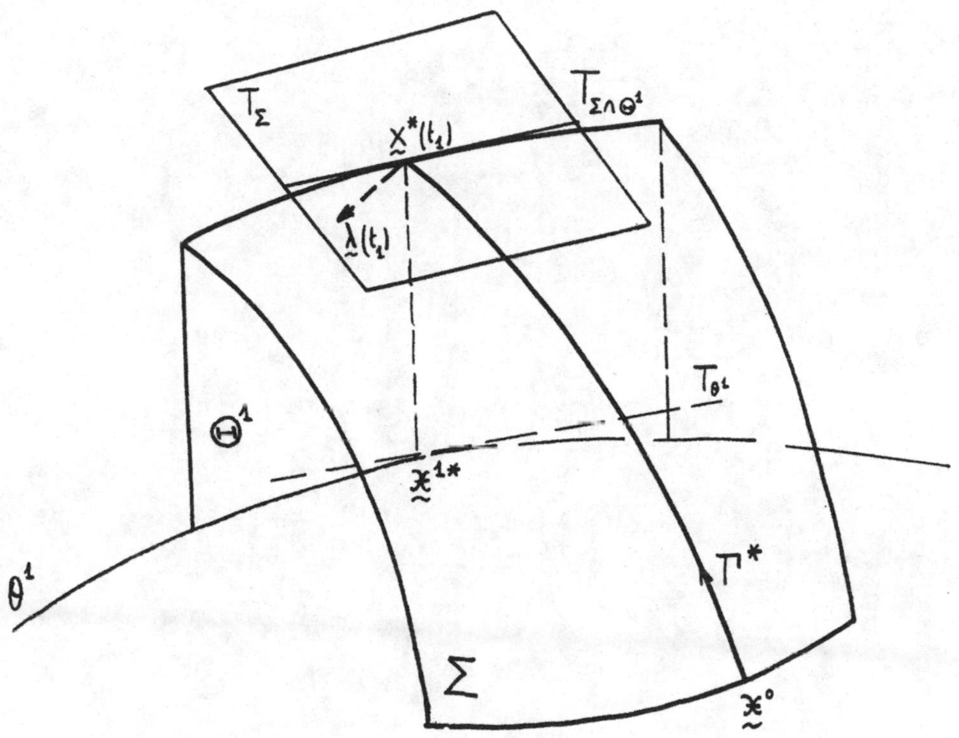

Fig. 1, Condition de transversalité terminale

B-point par rapport à $\sum_2$, (Fig. 2) . Alors , d'après les définitions des surfaces limites et des B-points, on a $C_1 < C_2$, et par conséquent

$$V^*(\underset{\sim}{\mathfrak{x}}^{o}_1 ; \theta^1) < V^*(\underset{\sim}{\mathfrak{x}}^{o}_2 ; \theta^1)$$

Ainsi, si la trajectoire optimale Γ^* est issue de $\underset{\sim}{x}^*(t_o) = (0, \underset{\sim}{\mathfrak{x}}^{o*})$, appartient à la surface limite $\sum$, et est optimale <u>par rapport à toutes les trajectoires dont les points initiaux appartiennent à</u> θ^o, c'est-à-dire si

$$V^*(\underset{\sim}{\mathfrak{x}}^{o*} ; \theta^1) \leqslant V^*(\underset{\sim}{\mathfrak{x}}^{o} ; \theta^1) \quad \forall \underset{\sim}{\mathfrak{x}}^{o} \in \theta^o \cap E^*$$

alors, tout point de $\theta^o \cap E^*$ appartient à la région $A / \sum \cup \sum$. Ceci implique que le plan tangent à θ^o en $\underset{\sim}{x}^*(t_o)$, $T_{\theta^o}(\underset{\sim}{x}^*(t_o))$ appartient au plan tangent à $\sum$, au point initial de Γ^*, $T_{\Sigma}(\underset{\sim}{x}^*(t_o))$, (fig. 3) 1) .

Compte tenu du choix de $\underset{\sim}{\lambda}(t_o)$, qui est normal à $T_{\Sigma}(\underset{\sim}{x}^*(t_o))$, on obtient la <u>condition de transversalité initiale</u>

$$\underset{\sim}{\lambda}(t_o) \cdot \eta^o = 0 \tag{30}$$

où $\eta^o = (\eta^o_o , \eta^o_1 , \ldots \eta^o_n)$ est un vecteur <u>quelconque</u> dans $T_{\theta^o}(\underset{\sim}{x}^*(t_o))$; c'est-à-dire

$$\eta^o_o = 0$$

$$\sum_{j=1}^{n} \left. \frac{\partial \theta^o_\rho(\underset{\sim}{\mathfrak{x}})}{\partial x_j} \right|_{\underset{\sim}{\mathfrak{x}} = \underset{\sim}{\mathfrak{x}}^{o*}} \eta_j = 0 \ , \quad \rho = 1, 2, \ldots p \tag{31}$$

On peut déterminer à partir de (31) p + 1 composantes

---

1) Rappelons que $T_{\theta^o} \subset E^n$ tandis que $T_{\Sigma} \subset E^{n+1}$

de η^o ; et les coefficients des n-p composantes arbitraires de η^o dans (30) doivent s'annuler. Comme $\lambda_o(t_o)$ n'apparait pas explicitement dans les n-p relations qui en résultent, la condition de transversalité initiale et les conditions initiales (29) fournissent un ensemble de n relations entre 2n variables.

13. Le Principe du Maximum pour les trajectoires optimales fortement régulières

Les résultats des paragraphes 10-12 s'appliquent au transfert, le long de trajectoires optimales fortement régulières, d'un système régi par les équations d'état (8) avec index de performance (9), entre une variété initiale donnée, θ^o, au temps t_o, et une variété terminale donnée, θ^1, au temps non spécifié t_1.

Posant

$$\mathcal{H}(\underset{\sim}{\lambda}, \underset{\sim}{x}, \underset{\sim}{u}) \triangleq \underset{\sim}{\lambda} \cdot \underset{\sim}{f}(\underset{\sim}{x}, \underset{\sim}{u})$$

nous pouvons énoncer le Théorème suivant :

THEOREME II. Si $\underset{\sim}{u}^*(t)$, $t_o \leqslant t \leqslant t_1$, est une commande optimale, pour laquelle la solution de (11) est la trajectoire optimale $\underset{\sim}{x}^*(t)$, il existe un vecteur non nul, fonction continue du temps, $\underset{\sim}{\lambda}(t)$, qui est une solution de (20) telle que

(i) $\sup_{\underset{\sim}{u} \in \mathcal{V}} \mathcal{H}(\underset{\sim}{\lambda}(t), \underset{\sim}{x}^*(t), \underset{\sim}{u}) = \mathcal{H}(\underset{\sim}{\lambda}(t), \underset{\sim}{x}^*(t), \underset{\sim}{u}^*(t))$;

(ii) $\mathcal{H}(\underset{\sim}{\lambda}(t), \underset{\sim}{x}^*(t), \underset{\sim}{u}^*(t)) = 0$;

(iii) $\lambda_o(t) = \text{constante} \leqslant 0$ quel que soit $t \in [t_o, t_1]$; et

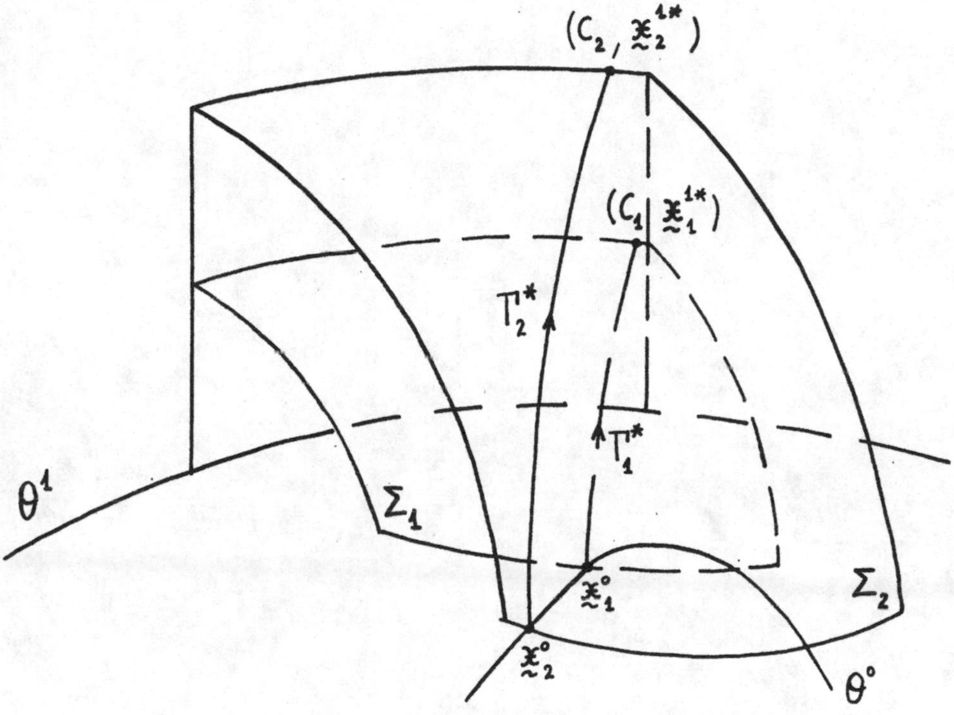

Fig. 2, Variété initiale

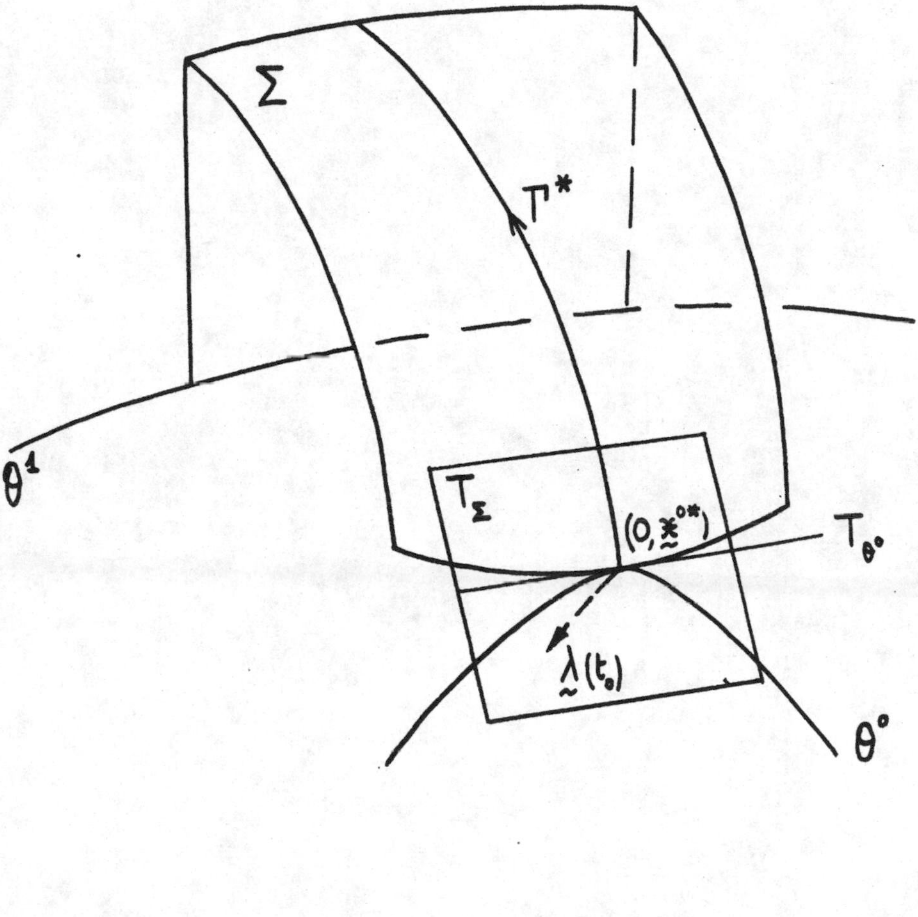

Fig. 3, Condition de transversalité initiale

(iv) le vecteur $(\lambda_1(t), \lambda_2(t), \ldots \lambda_n(t))$ est normal à θ° et θ^1 à $t = t_o$ et $t = t_1$, respectivement.

Ce Théorème, qui est le Principe du Maximum de Pontryagin[1], n'a été établi ici que dans le cas de points interieurs, fortement réguliers, des surfaces limites. La méthode géométrique exposée ici peut également être appliquée au cas de points interieurs faiblement réguliers ou non réguliers[2], ainsi qu'au cas de points frontières[3].

N.B. - La définition de points faiblement réguliers sera donnée dans la Partie II, paragraphe 11.

14. Equation de la Programmation Dynamique

Rappelons préalablement la définition d'une surface limite, Σ :

$$\Phi(\underset{\sim}{x}) \triangleq x_o + V^*(\underset{\sim}{x}; \theta^1) = C \tag{32}$$

et considérons

$$\operatorname{grad} \Phi(\underset{\sim}{x}) = \left(1, \frac{\partial V^*(\underset{\sim}{x}, \theta^1)}{\partial x_1}, \ldots \frac{\partial V^*(\underset{\sim}{x}; \theta^1)}{\partial x_n}\right) \tag{33}$$

1) L.S. Pontryagin et al., The Mathematical Theory of Optimal Processes, Interscience-Wiley, N.Y. 1962

2) G. Leitmann, Some Geometrical Aspects of Optimal Processes, SIAM Journal on Control, 3, 1965, pp. 53-65, et A. Blaquière, Further Investigation into the Geometry of Optimal Processes, SIAM Journal on Control, 4, 1966, pp. 19-33. Ce n'est qu'incidemment que le Principe du Maximum se trouve établi dans ces deux références, dont le principal objet est le développement de la théorie géométrique des Processus Optimaux.

3) A. Blaquière et G. Leitmann, On the Geometry of Optimal Processes-Part III, Report No. AM-66-1, Division of Applied Mechanics, Office of Naval Research, University of Cal. Berkeley.

Ce vecteur est défini au point $\underset{\sim}{x} = \underset{\sim}{x}^*(t)$ d'une trajectoire optimale Γ^*, pourvu que

(i) $\underset{\sim}{x}^*(t)$ soit un point interieur fortement régulier de la surface limite Σ qui contient Γ^* ;

(ii) $\partial V^*(\underset{\sim}{x}; \theta^1)/\partial x_j$, $\underset{\sim}{x} = \underset{\sim}{x}^*(t)$, $j = 1, 2, \ldots n$, soient définis.

Si la condition (i) est vérifiée, $\underset{\sim}{\lambda}(t)$ est normal à Σ et dirigé vers B/Σ. Si la condition (ii) est vérifiée , on a $n_o(\underset{\sim}{x}^*(t)) \neq 0$; par conséquent

$$\lambda_o(t) \equiv \text{constante} \neq 0$$

Comme toutes les relations qui sont utiles ici sont homogènes et de degré 1 par rapport aux variables adjointes, nous pouvons poser $\lambda_o(t) \equiv -1$

Comme grad $\Phi(\underset{\sim}{x}^*(t))$ et $\underset{\sim}{\lambda}(t)$, sont des vecteurs proportionnels, dont les composantes sur l'axe x_o sont 1 et -1, respectivement, nous avons

$$\underset{\sim}{\lambda}(t) = -\,\text{grad}\ \Phi(\underset{\sim}{x}^*(t)) \tag{34}$$

en tout point où les conditions (i) et (ii) sont vérifiées .

La relation (34) jointe aux propriétés (i) et (ii) du Théorème II, conduit à l'équation de la Programmation Dynamique

$$\underset{\underset{\sim}{u} \in U}{\text{Inf}} \left[f_o(\underset{\sim}{x}, \underset{\sim}{u}) + \sum_{j=1}^{n} f_j(\underset{\sim}{x}, \underset{\sim}{u}) \frac{\partial V^*(\underset{\sim}{x}; \theta^1)}{\partial x_j} \right]_{\underset{\sim}{x} = \underset{\sim}{x}^*(t)} = 0 \tag{35}$$

II

PROPRIETES LOCALES DES SURFACES LIMITES

1. Introduction

Dans la première partie de cet exposé, certaines propriétés globales des surfaces limites ont été mises en évidence, globales dans le sens qu'elles sont relatives au comportement des trajectoires et à leurs situations par rapport aux surfaces limites considérées dans leur ensemble. Ici nous étudierons quelques propriétés locales des surfaces limites, locales dans le sens qu'elles intéressent les voisinages des points d'une surface Σ donnée .

2. Une Hypothèse de Base

Dans la suite nous poserons $\widetilde{A/\Sigma} \triangleq (A/\Sigma) \cup \Sigma$ et $\widetilde{B/\Sigma} \triangleq (B/\Sigma) \cup \Sigma$ et nous introduirons une hypothèse de base qui concerne la façon dont une surface limite Σ donnée sépare $E^* \times x_o$ en deux régions disjointes. Ces régions seront $\widetilde{A/\Sigma}$ et B/Σ .

Soit η un vecteur borné en un point interieur $\underset{\sim}{x}$ d'une surface limite Σ donnée. Nous supposerons que, quel que soit η il existe un scalaire $\delta > 0$ tel que, quel que soit ε, $0 < \varepsilon < \delta$, le point $\underset{\sim}{x} + \varepsilon\eta$ appartient soit à la région $\widetilde{A/\Sigma}$, soit à la région B/Σ .

Cette hypothèse sera adoptée dans la suite.

3. Définition des Cones Locaux $\mathcal{C}_A(\underset{\sim}{x})$ et $\mathcal{C}_B(\underset{\sim}{x})$

A chaque point interieur d'une surface limite, nous associerons deux cones :

$$\mathcal{C}_A(\underset{\sim}{x}) \triangleq \left\{ \underset{\sim}{x} + \eta : \exists \alpha > 0 \text{ tel que } \forall \varepsilon,\ 0 < \varepsilon < \alpha,\ \underset{\sim}{x} + \varepsilon \eta \in \widetilde{A/\Sigma} \right\}$$

et

$$\mathcal{C}_B(\underset{\sim}{x}) \triangleq \left\{ \underset{\sim}{x} + \eta : \exists \beta > 0 \text{ tel que } \forall \varepsilon,\ 0 < \varepsilon < \beta,\ \underset{\sim}{x} + \varepsilon \eta \in B/\Sigma \right\}$$

η vecteur borné au point interieur $\underset{\sim}{x}$ de la surface limite Σ.

4. Points Interieurs de $\mathcal{C}_A(\underset{\sim}{x})$ et $\mathcal{C}_B(\underset{\sim}{x})$, une autre Hypothèse de Base

Avant d'introduire une deuxième hypothèse de base qui concerne également la façon dont une surface limite sépare E^{n+1}, nous définirons un point interieur d'un cone local $\mathcal{C}_A(\underset{\sim}{x})$ ou $\mathcal{C}_B(\underset{\sim}{x})$.

Nous dirons que $\underset{\sim}{x} + \eta$ est un point interieur de $\mathcal{C}_A(\underset{\sim}{x})$ (ou $\mathcal{C}_B(\underset{\sim}{x})$) si

(i) $\underset{\sim}{x} + \eta \in \mathcal{C}_A(\underset{\sim}{x})$ (ou $\mathcal{C}_B(\underset{\sim}{x})$) ; et

(ii) il existe une boule ouverte $B(\underset{\sim}{x} + \eta)$ dans E^{n+1} centrée en $\underset{\sim}{x} + \eta$ telle que tous les points de $B(\underset{\sim}{x} + \eta)$ appartiennent à $\mathcal{C}_A(\underset{\sim}{x})$ (ou $\mathcal{C}_B(\underset{\sim}{x})$).

Les définitions des ouvertures $\mathring{\mathcal{C}}_A(\underset{\sim}{x})$ et $\mathring{\mathcal{C}}_B(\underset{\sim}{x})$ s'ensuivent

$$\mathring{\mathcal{C}}_A(\underset{\sim}{x}) \triangleq \left\{ \underset{\sim}{x} + \eta : \underset{\sim}{x} + \eta \text{ est un point interieur de } \mathcal{C}_A(\underset{\sim}{x}) \right\}$$

$$\mathring{\mathcal{C}}_B(\underset{\sim}{x}) \triangleq \left\{ \underset{\sim}{x} + \eta : \underset{\sim}{x} + \eta \text{ est un point interieur de } \mathcal{C}_B(\underset{\sim}{x}) \right\}$$

Notre deuxième hypothèse de base est alors la suivante :

Soit $\underset{\sim}{x} + \eta'$ un point interieur de $\mathcal{C}_A(\underset{\sim}{x})$ (ou $\mathcal{C}_B(\underset{\sim}{x})$) ; c'est-à-dire qu'il existe une boule ouverte $B(\underset{\sim}{x} + \eta')$ dans E^{n+1}

qui appartient à $\mathscr{C}_A(\underset{\sim}{x})$ (ou $\mathscr{C}_B(\underset{\sim}{x})$). Alors il existe un boule ouverte $B'(\underset{\sim}{x}+\eta')$ dans E^{n+1} qui appartient à $\mathscr{C}_A(\underset{\sim}{x})$ (ou $\mathscr{C}_B(\underset{\sim}{x})$) et qui a la propriété suivante :

quel que soit $\underset{\sim}{x}+\eta$ dans $B'(\underset{\sim}{x}+\eta')$ il existe un nombre positif α (indépendant de η) tel que quel que soit ε, $0<\varepsilon\leqslant\alpha$, le point $\underset{\sim}{x}+\varepsilon\eta$ appartient à A/Σ (ou B/Σ)

En s'appuyant sur cette deuxième hypothèse de base, on peut établir le Lemme suivant

Lemme 1. Soit $\underset{\sim}{x}$ un point interieur d'une surface limite Σ, et $\underset{\sim}{x}+\eta'\in\mathring{\mathscr{C}}_A(\underset{\sim}{x})$, il existe une boule ouverte $B'(\underset{\sim}{x}+\eta')\subset\mathscr{C}_A(\underset{\sim}{x})$, et un nombre positif β tels que quel que soit $\underset{\sim}{x}+\eta\in B'(\underset{\sim}{x}+\eta')$, et quel que soit ε, $0<\varepsilon<\beta$, $\underset{\sim}{x}+\varepsilon\eta\in A/\Sigma$.

5. Cone Local $\mathscr{S}(\underset{\sim}{x})$

Considérons maintenant les fermetures $\bar{\mathscr{C}}_A(\underset{\sim}{x})$ et $\bar{\mathscr{C}}_B(\underset{\sim}{x})$, et définissons un autre cone local en $\underset{\sim}{x}$. Soit $\mathscr{S}(\underset{\sim}{x})$ ce cone :

$$\mathscr{S}(\underset{\sim}{x}) \triangleq \bar{\mathscr{C}}_A(\underset{\sim}{x}) \cap \bar{\mathscr{C}}_B(\underset{\sim}{x})$$

Nous verrons au paragraphe suivant que $\mathscr{S}(\underset{\sim}{x})$ n'est pas vide.

6. Une Partition de E^{n+1}

Nous établirons maintenant que

Lemme 2. Si $\underset{\sim}{x}$ est un point interieur d'une surface limite Σ,

<u>les cones locaux</u> $\mathcal{C}_A(\underset{\sim}{x})$ et $\mathcal{C}_B(\underset{\sim}{x})$ <u>constituent une partition de</u> E^{n+1} ; <u>en d'autres termes</u>

$$\mathcal{C}_A(\underset{\sim}{x}) = \text{comp}\ \mathcal{C}_B(\underset{\sim}{x}) \quad \underline{\text{et}} \quad \mathcal{C}_B(\underset{\sim}{x}) = \text{comp}\ \mathcal{C}_A(\underset{\sim}{x})$$

En premier lieu, remarquons que

$$\mathcal{C}_A(\underset{\sim}{x}) \cup \mathcal{C}_B(\underset{\sim}{x}) = E^{n+1}$$

Ceci est dû au fait que, quel que soit $\underset{\sim}{x} + \eta \in E^{n+1}$, ce point appartient soit à $\mathcal{C}_A(\underset{\sim}{x})$, soit à $\mathcal{C}_B(\underset{\sim}{x})$, qui sont les deux seules possibilités.

Pour établir le Lemme 2, il reste à prouver que ni $\mathcal{C}_A(\underset{\sim}{x})$ ni $\mathcal{C}_B(\underset{\sim}{x})$ n'est vide, et que l'intersection de ces deux cones est vide.

Considérons les demi-droites ouvertes L_- et L_+, issues de $\underset{\sim}{x}$, et parallèles à l'axe x_0 respectivement dans le sens négatif et dans le sens positif. Comme conséquence des définitions de A/Σ, B/Σ et $A\widetilde{/}\Sigma$, nous avons

$$\forall\, \underset{\sim}{x} + \varepsilon\eta \in L_+ , \quad \underset{\sim}{x} + \varepsilon\eta \in A\widetilde{/}\Sigma$$

$$\forall\, \underset{\sim}{x} + \varepsilon\eta \in L_- , \quad \underset{\sim}{x} + \varepsilon\eta \in B/\Sigma$$

Ainsi, il résulte des définitions de $\mathcal{C}_A(\underset{\sim}{x})$ et $\mathcal{C}_B(\underset{\sim}{x})$ que

$$\mathcal{C}_A(\underset{\sim}{x}) \neq \emptyset \quad , \quad \mathcal{C}_B(\underset{\sim}{x}) \neq \emptyset$$

Finalement nous établirons que

(1) $$\mathcal{C}_A(\underset{\sim}{x}) \cap \mathcal{C}_B(\underset{\sim}{x}) = \emptyset$$

Pour prouver cette propriété, supposons qu'elle est

incorrecte; c'est-à-dire que

$$\mathscr{C}_A(\underset{\sim}{x}) \cap \mathscr{C}_B(\underset{\sim}{x}) \neq \emptyset$$

et considérons un point $\underset{\sim}{x} + \eta$ tel que

$$\underset{\sim}{x} + \eta \in \mathscr{C}_A(\underset{\sim}{x}) \qquad \text{et} \qquad \underset{\sim}{x} + \eta \in \mathscr{C}_B(\underset{\sim}{x})$$

Alors

$$\exists \alpha > 0 \text{ tel que } \forall \varepsilon \,,\quad 0 < \varepsilon < \alpha \quad ; \quad \underset{\sim}{x} + \varepsilon\eta \in \widetilde{A/\Sigma}$$

$$\exists \beta > 0 \text{ tel que } \forall \varepsilon \,,\quad 0 < \varepsilon < \beta \quad , \quad \underset{\sim}{x} + \varepsilon\eta \in B/\Sigma$$

Supposons, par exemple, que $\alpha < \beta$; alors

$\forall \varepsilon$, $0 < \varepsilon < \alpha$, $\underset{\sim}{x} + \varepsilon\eta \in \widetilde{A/\Sigma}$ et $\underset{\sim}{x} + \varepsilon\eta \in B/\Sigma$ ce qui implique

$$(\widetilde{A/\Sigma}) \cap (B/\Sigma) \neq \emptyset$$

Ceci est incompatible avec les propriétés de $\widetilde{A/\Sigma}$ et B/Σ ; donc (1) est établi et par conséquent le Lemme 2.

Comme conséquence directe du Lemme 2 :

<u>Corollaire 1.</u> <u>Le cone local $\mathscr{S}(\underset{\sim}{x}) \triangleq \overline{\mathscr{C}}_A(\underset{\sim}{x}) \cap \overline{\mathscr{C}}_B(\underset{\sim}{x})$ est la frontière commune des cones locaux $\mathscr{C}_A(\underset{\sim}{x})$ et $\mathscr{C}_B(\underset{\sim}{x})$</u>

On sait que

$$\partial \mathscr{C}_A(\underset{\sim}{x}) = \overline{\mathscr{C}}_A(\underset{\sim}{x}) \cap \overline{\text{comp } \mathscr{C}_A(\underset{\sim}{x})}$$

$$\partial \mathscr{C}_B(\underset{\sim}{x}) = \overline{\mathscr{C}}_B(\underset{\sim}{x}) \cap \overline{\text{comp } \mathscr{C}_B(\underset{\sim}{x})}$$

Ces relations, jointes au Lemme 2, établissent le Corollaire 1.

De plus :

Corollaire 2. Les cones locaux $\mathcal{C}_A(\underset{\sim}{x})$ et $\mathcal{C}_B(\underset{\sim}{x})$ vérifient les relations suivantes;

$$\text{comp}\ \overset{\circ}{\mathcal{C}}_A(\underset{\sim}{x}) = \overline{\mathcal{C}}_B(\underset{\sim}{x})$$
$$\text{comp}\ \overset{\circ}{\mathcal{C}}_B(\underset{\sim}{x}) = \overline{\mathcal{C}}_A(\underset{\sim}{x})$$

Ce corollaire est une conséquence du Lemme 2, avec

$$\overline{\text{comp}\ \mathcal{C}_A(\underset{\sim}{x})} = \text{comp}\ \overset{\circ}{\mathcal{C}}_A(\underset{\sim}{x})$$
$$\overline{\text{comp}\ \mathcal{C}_B(\underset{\sim}{x})} = \text{comp}\ \overset{\circ}{\mathcal{C}}_B(\underset{\sim}{x})$$

7. Lemmes 3 et 4

Nous établirons ensuite les Lemmes suivants

Lemme 3. Soit η un vecteur borné en un point interieur $\underset{\sim}{x}$ d'une surface limite Σ. Si $\eta = \eta(\varepsilon)$ est une fonction continue du parametre ε, avec

(i) $\eta(\varepsilon) \to \underset{\sim}{l}$ quand $\varepsilon \to 0$

(ii) $\exists\, \gamma > 0$ tel que $\forall \varepsilon,\ 0 < \varepsilon < \gamma,\ \underset{\sim}{x} + \varepsilon \eta(\varepsilon) \in \widetilde{A/\Sigma}$

alors

$$\underset{\sim}{x} + \underset{\sim}{l} \in \overline{\mathcal{C}}_A(\underset{\sim}{x})$$

Pour établir ce Lemme, supposons qu'il est incorrect, c'est-à-dire que

$$\underset{\sim}{x} + \underset{\sim}{l} \notin \overline{\mathcal{C}}_A(\underset{\sim}{x})$$

On a, d'après le Corollaire 2

$$\underset{\sim}{x} + \underset{\sim}{l} \in \overset{\circ}{\mathcal{C}}_B(\underset{\sim}{x})$$

Alors, d'après la deuxieme hypothèse de base , il existe une boule ouverte $B'(\underset{\sim}{x} + \underset{\sim}{\ell})$ et un nombre positif α tels que, quell que soit $\underset{\sim}{x} + \eta \in B'(\underset{\sim}{x} + \underset{\sim}{\ell})$, et quel que soit ε, $0 < \varepsilon < \alpha$, le point $\underset{\sim}{x} + \varepsilon\eta$ appartient à B/Σ.

Cependant, puisque $\eta(\varepsilon) \rightarrow \underset{\sim}{\ell}$ quand $\varepsilon \rightarrow 0$, il existe un nombre positif $\delta < \gamma$ tel que pour ε, $0 < \varepsilon < \delta$, le point $\underset{\sim}{x} + \eta(\varepsilon)$ appartient à $B'(\underset{\sim}{x} + \underset{\sim}{\ell})$. Par conséquent, il existe un nombre positif β tel que quel que soit ε, $0 < \varepsilon < \beta$,

$$\underset{\sim}{x} + \varepsilon\eta(\varepsilon) \in B/\Sigma$$

ce qui est en contradiction avec (ii) du Lemme 3. Le Lemme 3 est donc établi.

On établirait le Lemme 4 de façon analogue :

Lemme 4. Soit η un vecteur borné en un point interieur $\underset{\sim}{x}$ d'une surface limite Σ. Si $\eta = \eta(\varepsilon)$ est une fonction continue du paramètre ε, avec

(i) $\eta(\varepsilon) \rightarrow \underset{\sim}{\ell}$ quand $\varepsilon \rightarrow 0$

(ii) $\exists \gamma > 0$ tel que $\forall \varepsilon$, $0 < \varepsilon < \gamma$, $\underset{\sim}{x} + \varepsilon\eta(\varepsilon) \in \widetilde{B/\Sigma}$

alors

$$\underset{\sim}{x} + \underset{\sim}{\ell} \in \overline{\mathcal{C}}_B(x)$$

Comme conséquence de ces Lemmes, nous avons

Corollaire 3. Soit η' un vecteur borné en un point interieur $\underset{\sim}{x}$ d'une surface limite Σ. Si $\eta = \eta(\varepsilon)$ est une fonction continue du paramètre ε, avec

(i) $\eta(\varepsilon) \to \underset{\sim}{\ell}$ quand $\varepsilon \to 0$

(ii) $\exists \gamma > 0$ tel que $\forall \varepsilon,\ 0 < \varepsilon < \gamma\ ;\ \underset{\sim}{x} + \varepsilon \eta(\varepsilon) \in \Sigma$

alors

$$\underset{\sim}{x} + \underset{\sim}{\ell} \in \mathcal{S}(\underset{\sim}{x})$$

8. Cone des Vecteurs f

Dans la suite nous considérerons des systèmes régis par

$$\text{(2)} \qquad \frac{dx_j}{dt} = f_j\,(x_1, \ldots x_n,\ u_1, \ldots u_m) \quad j = 1, \ldots n$$

où $u_1, \ldots u_m$ sont les variables de commande.

Lorsqu'on spécifie les fonctions du temps $u_1(t), \ldots u_m(t)$, $t_o \leqslant t \leqslant t_1$, ces équations définissent les règles auxquelles obéit le système sur l'intervalle de temps $[t_o, t_1]$.

Nous supposerons que le vecteur $\underset{\sim}{u}$, de composantes $u_1, \ldots u_m$, appartient à un sous - ensemble U de E^m.

De plus, suivant les hypothèses introduites dans la première partie de l'exposé nous considérerons la fonctionnelle

$$\int_{t_o}^{t_1} f_o\,\big(\underset{\sim}{x}(t),\ \underset{\sim}{u}(t)\big)\,dt$$

où la fonction de commande $\underset{\sim}{u}(t)$ transfère le système, dans E^n, de l'état initial $\underset{\sim}{x}^o$ à un état final, dans le temps non spécifié $t_1 - t_o$.

On peut alors établir aisément que l'équation de Γ , ou $\overline{\Gamma}$, est

$$x_o(t) + \int_t^{t_1} f_o \left(\underset{\sim}{x}(\tau), \underset{\sim}{u}(\tau) \right) d\tau = C$$

D'où

$$(3) \qquad \frac{dx_o}{dt} = f_o(\underset{\sim}{x}, \underset{\sim}{u})$$

Si nous réunissons (2) et (3) en une seule équation vectorielle, nous avons

$$(4) \qquad \frac{d\underset{\sim}{x}}{dt} = f(\underset{\sim}{x}, \underset{\sim}{u})$$

où

$$\underset{\sim}{f}(\underset{\sim}{x}, \underset{\sim}{u}) = \left((f_o(\underset{\sim}{x}, \underset{\sim}{u}), f_1(\underset{\sim}{x}, \underset{\sim}{u}), \ldots f_n(\underset{\sim}{x}, \underset{\sim}{u}) \right)$$

De plus nous supposerons que les fonctions

$$f_j(\underset{\sim}{x}, \underset{\sim}{u}) \quad \text{et} \quad \partial f_j(\underset{\sim}{x}, \underset{\sim}{u}) / \partial x_i$$

$$i, j = 1, 2, \ldots n$$

sont définies et continues sur $E^n \times \mathcal{U}$.

Par conséquent, si nous nous donnons un vecteur constant $\underset{\sim}{u}^b \in \mathcal{U}$, (4) définit un champ de vecteurs constant , c'est-à-dire un champ de vecteurs vitesses dans E^{n+1} , qui a les propriétés suivantes

(i) Les lignes de force du champ sont les courbes intégrales de l'équation (4) ;

(ii) Par chaque point de E^{n+1} passe une telle trajectoire et une seule, dont la tangente est uniquement définie en ce point.

Considérons maintenant une telle courbe intégrale L dont le point courant passe par $\underset{\sim}{x} \in \Sigma$ au temps t, qui peut être arbitrairement choisi, et soit

$$\eta_+(\Delta t) \triangleq \frac{\Delta \underset{\sim}{x}_+}{\Delta t} \quad , \quad \eta_-(\Delta t) \triangleq \frac{\Delta \underset{\sim}{x}_-}{\Delta t}$$

où

$$\Delta \underset{\sim}{x}_+ \triangleq \underset{\sim}{x}(t_i + \Delta t) - \underset{\sim}{x}(t_i)$$

$$\Delta \underset{\sim}{x}_- \triangleq \underset{\sim}{x}(t_i - \Delta t) - \underset{\sim}{x}(t_i)$$

$$\underset{\sim}{x}(t_i) = \underset{\sim}{x} \quad , \quad \Delta t > 0 \quad , \quad t_i \in (t', t'')$$

$\eta_+(\Delta t)$ et $\eta_-(\Delta t)$ sont des fonctions continues de Δt,

et

$$\eta_+(\Delta t) \longrightarrow \underset{\sim}{f}(\underset{\sim}{x}, \underset{\sim}{u}^b)$$

$$\eta_-(\Delta t) \longrightarrow -\underset{\sim}{f}(\underset{\sim}{x}, \underset{\sim}{u}^b)$$

quand $\Delta t \rightarrow 0$

Comme conséquence des propriétés globales des surfaces Σ, nous avons

$$\underset{\sim}{x} + \Delta \underset{\sim}{x}_+ = \underset{\sim}{x} + \eta_+(\Delta t)\,\Delta t \in A/\widetilde{\Sigma}$$

$$\underset{\sim}{x} + \Delta \underset{\sim}{x}_- = \underset{\sim}{x} + \eta_-(\Delta t)\,\Delta t \in B/\widetilde{\Sigma}$$

pour Δt suffisemment petit.

On deduit alors des Lemmes 3 et 4

Lemme 5. En un point interieur $\underset{\sim}{x}$ d'une surface limite Σ

$$\underset{\sim}{x} + \underset{\sim}{f}(\underset{\sim}{x}, \underset{\sim}{u}) \in \overline{\mathcal{C}}_A(\underset{\sim}{x})$$

$$\underset{\sim}{x} - \underset{\sim}{f}(\underset{\sim}{x}, \underset{\sim}{u}) \in \overline{\mathcal{C}}_B(\underset{\sim}{x})$$

quel que soit $\underset{\sim}{u}$, $\underset{\sim}{u} \in \mathcal{U}$.

9. Hyperplan Séparant d'un Cone Local

Introduisons ici quelques définitions.

Nous dirons qu'un hyperplan à n dimensions, $\mathcal{J}(\underset{\sim}{x})$, en un point interieur $\underset{\sim}{x}$ d'une surface limite Σ, est un hyperplan séparant du cone fermé $\overline{\mathcal{C}}_A(\underset{\sim}{x})$ (ou $\overline{\mathcal{C}}_B(\underset{\sim}{x})$) ; si tout point

$$\underset{\sim}{x} + \eta \in \overline{\mathcal{C}}_A(\underset{\sim}{x}) \qquad (\text{ou } \overline{\mathcal{C}}_B(\underset{\sim}{x}))$$

appartient à l'un des demi-espaces fermés déterminés par $\mathcal{J}(\underset{\sim}{x})$.

Le demi-espace fermé correspondant sera désigné par $\overline{R}_A$ (ou $\overline{R}_B$) , et le demi-espace ouvert correspondant par R_A (ou R_B) .

Alors , si le cone fermé $\overline{\mathcal{C}}_A(\underset{\sim}{x})$ (ou $\overline{\mathcal{C}}_B(\underset{\sim}{x})$) admet un hyperplan séparant à n dimensions, nous dirons que ce cone $\overline{\mathcal{C}}_A(\underset{\sim}{x})$ (ou $\overline{\mathcal{C}}_B(\underset{\sim}{x})$) est séparable.

Evidemment, si $\mathcal{J}(\underset{\sim}{x})$ est un hyperplan séparant à n dimension de $\overline{\mathcal{C}}_A(\underset{\sim}{x})$ ou $\overline{\mathcal{C}}_B(\underset{\sim}{x})$, respectivement, on a

$$\text{(5)} \qquad \overline{\mathcal{C}}_A(\underset{\sim}{x}) \subseteq \overline{R}_A \Longrightarrow \begin{cases} \mathcal{C}_A(\underset{\sim}{x}) \subseteq \overline{R}_A \\ \mathring{\mathcal{C}}_A(\underset{\sim}{x}) \subseteq R_A \end{cases}$$

$$(6) \qquad \overline{\mathcal{C}}_B(\underset{\sim}{x}) \subseteq \overline{R}_B \Rightarrow \begin{cases} \mathcal{C}_B(\underset{\sim}{x}) \subseteq \overline{R}_B \\ \mathring{\mathcal{C}}_B(\underset{\sim}{x}) \subseteq R_B \end{cases}$$

10. Cone des Normales

Si $\overline{\mathcal{C}}_A(\underset{\sim}{x})$ (ou $\overline{\mathcal{C}}_B(\underset{\sim}{x})$) est séparable, considérons un hyperplan séparant $\mathcal{J}(\underset{\sim}{x})$ et un vecteur borné $\underset{\sim}{n}(\underset{\sim}{x})$, $|\underset{\sim}{n}(\underset{\sim}{x})| = 1$, au point $\underset{\sim}{x}$, normal à $\mathcal{J}(\underset{\sim}{x})$. De plus

(i) si $\overline{\mathcal{C}}_A(\underset{\sim}{x})$ est séparable, nous choisirons $\underset{\sim}{n}(\underset{\sim}{x})$ de telle sorte que

$$\underset{\sim}{x} + \underset{\sim}{n}(\underset{\sim}{x}) \in \text{comp} \quad \overline{R}_A$$

(ii) si $\overline{\mathcal{C}}_B(\underset{\sim}{x})$ est séparable, nous choisirons $\underset{\sim}{n}(\underset{\sim}{x})$ de telle sorte que

$$\underset{\sim}{x} + \underset{\sim}{n}(\underset{\sim}{x}) \in R_B$$

Remarquons que (i) et (ii) sont équivalants si $\overline{\mathcal{C}}_A(\underset{\sim}{x})$ et $\overline{\mathcal{C}}_B(\underset{\sim}{x})$ sont tous deux séparables.

Finalement, si $\overline{\mathcal{C}}_A(\underset{\sim}{x})$ (ou $\overline{\mathcal{C}}_B(\underset{\sim}{x})$) est séparable, nous considèrerons l'ensemble $\{\underset{\sim}{n}(\underset{\sim}{x})\}$ des vecteurs $\underset{\sim}{n}(\underset{\sim}{x})$ définis ci-dessus - en utilisant (i) ou (ii) suivant le cas - associés à tous les plans séparants de $\overline{\mathcal{C}}_A(\underset{\sim}{x})$ (ou $\overline{\mathcal{C}}_B(\underset{\sim}{x})$), et nous définirons le <u>cone des normales</u> au point $\underset{\sim}{x}$ par

$$\mathcal{C}_n(\underset{\sim}{x}) \triangleq \left\{ \underset{\sim}{x} + k\,\underset{\sim}{n}(\underset{\sim}{x}) : \; k > 0 \,, \; n(\underset{\sim}{x}) \in \{\underset{\sim}{n}(\underset{\sim}{x})\} \right\}$$

11. Points Interieurs Faiblement Réguliers, et Nonréguliers, d'une Surface Limite

Nous dirons qu'un point interieur $\underset{\sim}{x}$ d'une surface limite Σ est un point interieur faiblement régulier de Σ, si $\overline{\mathscr{C}}_A(\underset{\sim}{x})$ et $\overline{\mathscr{C}}_B(\underset{\sim}{x})$ sont tous deux séparables. Il est clair qu'un point interieur fortement régulier (I, 8) de Σ est un point faiblement régulier de Σ, mais la réciproque n'est pas vraie. Dans la suite nous appellerons simplement point régulier un point faiblement régulier.

Pour préciser cette notion, supposons que $\underset{\sim}{x} \in \Sigma$ est un point de cette espèce, et soit $\mathfrak{J}_A(\underset{\sim}{x})$ un plan séparant de $\overline{\mathscr{C}}_A(\underset{\sim}{x})$, par exemple. $\overline{R}_A$ est le demi-espace fermé déterminé par $\mathfrak{J}_A(\underset{\sim}{x})$, qui contient $\overline{\mathscr{C}}_A(\underset{\sim}{x})$. D'après (5) nous avons

$$\mathscr{C}_A(\underset{\sim}{x}) \subseteq \overline{R}_A$$

Mais

$$\mathscr{C}_B(\underset{\sim}{x}) = \text{comp}\ \mathscr{C}_A(\underset{\sim}{x})$$

De telle sorte que

$$(7) \qquad \begin{aligned} \mathscr{C}_A(\underset{\sim}{x}) \subseteq \overline{R}_A &\Rightarrow \text{comp}\ \overline{R}_A \subseteq \mathscr{C}_B(\underset{\sim}{x}) \\ &\Rightarrow \text{comp}\ R_A \subseteq \overline{\mathscr{C}}_B(\underset{\sim}{x}) \end{aligned}$$

Maintenant désignons par $\mathfrak{J}_B(\underset{\sim}{x})$ un hyperplan séparant à n dimensions de $\overline{\mathscr{C}}_B(\underset{\sim}{x})$, et par $\overline{R}_B$ le demi-espace fermé qu'il détermine et qui contient $\overline{\mathscr{C}}_B(\underset{\sim}{x})$.

On a

$$\overline{\mathscr{C}}_B(\underset{\sim}{x}) \subseteq \overline{R}_B$$

et d'après (6) et (7) :

(8) $$\text{comp } R_A \subseteq \overline{\mathscr{C}}_B(\underset{\sim}{x}) \subseteq \overline{R}_B$$

Comme les plans $\mathfrak{I}_A(\underset{\sim}{x})$ et $\mathfrak{I}_B(\underset{\sim}{x})$, frontières des demi-espaces fermés comp R_A et $\overline{R}_B$, respectivement, passent tous deux par $\underset{\sim}{x}$, (8) implique

(9) $$\text{comp } R_A = \overline{R}_B$$

ce qui implique d'ailleurs

$$\mathfrak{I}_A(\underset{\sim}{x}) = \mathfrak{I}_B(\underset{\sim}{x}) \triangleq \mathfrak{I}(\underset{\sim}{x})$$

De plus, (8) implique

(10) $$\overline{\mathscr{C}}_B(\underset{\sim}{x}) = \overline{R}_B$$

D'autre part, comme

$$\mathring{\mathscr{C}}_A(\underset{\sim}{x}) = \text{comp } \overline{\mathscr{C}}_B(\underset{\sim}{x})$$

il résulte de (9) et (10) que

$$\mathring{\mathscr{C}}_A(\underset{\sim}{x}) = R_A$$

d'où (*)

(11) $$\overline{\mathscr{C}}_A(\underset{\sim}{x}) = \overline{R}_A$$

(*) En effet $\overline{\mathring{\mathscr{C}}}_A(\underset{\sim}{x}) \subseteq \overline{\mathscr{C}}_A(\underset{\sim}{x})$, et on a d'autre part $\overline{\mathring{\mathscr{C}}}_A(\underset{\sim}{x}) = \overline{R}_A$ et $\overline{\mathscr{C}}_A(\underset{\sim}{x}) \subseteq \overline{R}_A$.

Finalement, d'après (10) et (11) nous avons

$$\mathcal{S}(\underset{\sim}{x}) \triangleq \bar{\ell}_A(\underset{\sim}{x}) \cap \bar{\ell}_B(\underset{\sim}{x}) = \bar{R}_A \cap \bar{R}_B = \mathcal{J}(\underset{\sim}{x})$$

Ces conclusions sont importantes car elles mettent en lumière un caractère fondamental des points interieurs réguliers de Σ : $\bar{\mathcal{C}}_A(\underset{\sim}{x})$ et $\bar{\mathcal{C}}_B(\underset{\sim}{x})$ ont le même hyperplan séparant

$$\mathcal{J}(\underset{\sim}{x}) = \mathcal{S}(\underset{\sim}{x})$$

et, de plus, cet hyperplan séparant est unique, puisque $\mathcal{S}(\underset{\sim}{x})$ est unique.

D'autre part si $\bar{\mathcal{C}}_A(\underset{\sim}{x})$ et $\bar{\mathcal{C}}_B(\underset{\sim}{x})$ ne sont pas tous deux séparables, le point $\underset{\sim}{x}$ est dit point interieur nonrégulier de Σ.

Différents cas doivent alors être envisagés :

(i) $\bar{\mathcal{C}}_A(\underset{\sim}{x})$ est séparable, et $\bar{\mathcal{C}}_B(\underset{\sim}{x})$ non séparable.

(ii) $\bar{\mathcal{C}}_B(\underset{\sim}{x})$ est séparable, et $\bar{\mathcal{C}}_A(\underset{\sim}{x})$ non séparable.

(iii) Ni $\bar{\mathcal{C}}_A(\underset{\sim}{x})$ ni $\bar{\mathcal{C}}_B(\underset{\sim}{x})$ n'est séparable; dans ce cas nous dirons que $\underset{\sim}{x}$ est un point antirégulier de Σ.

12. Transformation Linéaire (Rappel)

A chaque trajectoire optimale Γ^*, de fonction de commande $\underset{\sim}{u}^*(t)$, $t_o \leqslant t \leqslant t_1$, nous avons associé une transformation linéaire, définie par l'équation variationnelle

$$(12) \qquad \dot{\eta} = \left(\frac{\partial \underset{\sim}{f}}{\partial \underset{\sim}{x}}\right)\Big|_{\underset{\sim}{x} = x^*(t)} \eta$$

L'équation (12) définit un opérateur linéaire nonsingulier $A(t', t)$ tel que

$$\underset{\sim}{\eta}(t) = A(t', t)\,\underset{\sim}{\eta}' \qquad t' \leqslant t \leqslant t_1$$

où $\underset{\sim}{\eta}(t)$ est défini au point $\underset{\sim}{x}^*(t)$, et $\underset{\sim}{\eta}' \triangleq \underset{\sim}{\eta}(t')$ au point $\underset{\sim}{x}^*(t')$.

Certaines propriétés de cet opérateur linéaire ont été précisées dans la première partie de cet exposé. Parmi les plus importantes :

(i) L'équation adjointe associée à (12) est

$$(13) \qquad \overset{\circ}{\underset{\sim}{\lambda}} = -\left(\frac{\partial \underset{\sim}{f}}{\partial \underset{\sim}{x}}\right)^T \Bigg|_{\underset{\sim}{x} = \underset{\sim}{x}^*(t)} \underset{\sim}{\lambda}$$

Quand la condition initiale $\underset{\sim}{\lambda}(t_o) = \underset{\sim}{\lambda}^o$ est spécifiée, la solution de (13) est unique et continue sur $[t_o, t_1]$.

$$(14) \quad (ii) \qquad \underset{\sim}{\lambda}(t) \cdot \underset{\sim}{\eta}(t) = \text{constante} \qquad t_o \leqslant t \leqslant t_1$$

<u>13. Lemmes 6 et 7</u>

Ici nous nous bornerons à énoncer les deux Lemmes importants suivants

<u>Lemme 6. Soient $\underset{\sim}{x}^*(t')$ et $\underset{\sim}{x}^*(t'')$, $t_o \leqslant t' \leqslant t'' \leqslant t_1$ deux points de la trajectoire optimale Γ^*, correspondant à la fonction de commande $\underset{\sim}{u}^*(t)$ et à la solution $\underset{\sim}{x}^*(t)$, $t_o \leqslant t \leqslant t_1$. Soit $\underset{\sim}{\eta}'$ un vecteur en $\underset{\sim}{x}^*(t')$, et $\underset{\sim}{\eta}''$ son transformé en $\underset{\sim}{x}^*(t'')$;</u>

$$(15) \qquad \underset{\sim}{\eta}'' = A(t', t'')\,\underset{\sim}{\eta}'$$

<u>Si</u>

$$\underset{\sim}{x}^*(t') + \underset{\sim}{\eta}' \in \overline{\mathcal{C}}_A(\underset{\sim}{x}^*(t'))$$

alors

$$\underset{\sim}{x}^*(t'') + \eta'' \in \overline{\mathcal{C}}_A(\underset{\sim}{x}^*(t''))$$

Lemme 7. Soient $\underset{\sim}{x}^*(t')$ et $\underset{\sim}{x}^*(t'')$, $t_o \leqslant t' \leqslant t'' \leqslant t_1$, deux points de la trajectoire optimale Γ^*, correspondant à la fonction de commande $\underset{\sim}{u}^*(t)$ et à la solution $\underset{\sim}{x}^*(t)$, $t_o \leqslant t \leqslant t_1$. Soit η'' un vecteur en $\underset{\sim}{x}^*(t'')$ tel que

$$\underset{\sim}{x}^*(t'') + \eta'' \in \overline{\mathcal{C}}_B(\underset{\sim}{x}^*(t''))$$

η'' est le transformé , par la transformation linéaire (15), d'un vecteur η' en $\underset{\sim}{x}^*(t')$ tel que

$$\underset{\sim}{x}^*(t') + \eta' \in \mathcal{C}_B(\underset{\sim}{x}^*(t'))$$

14. Théorèmes de Séparabilité

Deux Théorèmes jouent un role particulièrement important dans la théorie géométrique des processus optimaux , ce sont

Théorème I. Si $\overline{\mathcal{C}}_B(\underset{\sim}{x}^*(t'))$ est séparable, $\overline{\mathcal{C}}_B(\underset{\sim}{x}^*(t''))$ est séparable.

Théorème II. Si $\overline{\mathcal{C}}_A(\underset{\sim}{x}^*(t''))$ est séparable $\overline{\mathcal{C}}_A(\underset{\sim}{x}^*(t'))$ est séparable.

Pour établir le Théorème I, considérons un hyperplan séparant à n dimensions, $\mathcal{J}(\underset{\sim}{x}^*(t'))$, de $\overline{\mathcal{C}}_B(\underset{\sim}{x}^*(t'))$. Ce plan passe par le point $\underset{\sim}{x}^*(t')$, et détermine le demi-espace fermé $\overline{R'}_B$, $\overline{\mathcal{C}}_B(\underset{\sim}{x}^*(t')) \subseteq \overline{R'}_B$ (16) , et le demi-espace ouvert comp $\overline{R'}_B$

On a

$$\underset{\sim}{x}^*(t') + \eta' \in \bar{\mathcal{C}}_B(\underset{\sim}{x}^*(t')) \Longrightarrow \underset{\sim}{x}^*(t') + \eta' \in \bar{R}'_B$$

Désignons par $\Pi(\underset{\sim}{x}^*((t''))$ et P''_B les transformés, par la transformation linéaire (15), de $\mathcal{J}(\underset{\sim}{x}^*(t'))$ et R'_B, respectivement. Comme la transformation $A(t', t'')$ est linéaire et nonsingulière, on peut voir aisément que

(i) $\Pi(\underset{\sim}{x}^*(t''))$ est la frontière commune de P''_B et comp P'_B, c'est-à-dire

$$\Pi(\underset{\sim}{x}^*(t'')) = \bar{P}''_B \cap \overline{\text{comp } \bar{P}''_B}$$

(ii) le transformé de comp $\bar{R}'_B$ est comp $\bar{P}''_B$.

Considérons maintenant un vecteur η'' en $\underset{\sim}{x}^*(t'')$, tel que

(17) $$\underset{\sim}{x}^*(t'') + \eta'' \in \bar{\mathcal{C}}_B(\underset{\sim}{x}^*(t''))$$

et supposons que

$$\underset{\sim}{x}^*(t'') + \eta'' \in \text{comp } \bar{P}''_B$$

D'après (ii) c'est le transformé d'un vecteur η' en $\underset{\sim}{x}^*(t')$, tel que

(18) $$\underset{\sim}{x}^*(t') + \eta' \in \text{comp } \bar{R}'_B$$

De plus, d'après, (17) et le Lemme 7, nous avons

(19) $$\underset{\sim}{x}^*(t') + \eta' \in \bar{\mathcal{C}}_B(\underset{\sim}{x}^*(t'))$$

Cependant, (19) joint à (16) est incompatible avec (18).

Nous concluons donc que

$$\underset{\sim}{x}^*(t'') + \eta'' \in \bar{P}''_B$$

quel que soit η'' vérifiant (17)

Finalement :

(i) $\bar{\mathcal{C}}_B(\underset{\sim}{x}^*(t''))$ est séparable ;

(ii) $\Pi(\underset{\sim}{x}^*(t''))$ est un plan séparant $\mathcal{T}(\underset{\sim}{x}^*(t''))$ de $\mathcal{C}_B(\underset{\sim}{x}^*(t''))$; et

(iii) $\bar{\mathcal{C}}_B(\underset{\sim}{x}^*(t''))$ appartient à $\bar{P}''_B$ qui est le transformé de $\bar{R}'_B$. Nous désignerons donc P''_B par R''_B.

Ainsi le Théorème I est établi. Un raisonnement analogue permettrait de prouver le Théorème II.

15. Corollaires 4 et 5

Les Théorèmes I et II ont les Corollaires suivants

Corollaire 4. Une trajectoire optimale Γ^* ne peut relier les points $\underset{\sim}{x}' = \underset{\sim}{x}^*(t')$ et $\underset{\sim}{x}'' = \underset{\sim}{x}^*(t'')$, $t'' > t'$, si $\bar{\mathcal{C}}_B(\underset{\sim}{x}')$ est séparable et $\bar{\mathcal{C}}_B(\underset{\sim}{x}'')$ nonséparable.

Corollaire 5. Une trajectoire optimale Γ^* ne peut relier les points $\underset{\sim}{x}' = \underset{\sim}{x}^*(t')$ et $\underset{\sim}{x}'' = \underset{\sim}{x}^*(t'')$, $t'' > t'$, si $\bar{\mathcal{C}}_A(\underset{\sim}{x}')$ est nonséparable et $\bar{\mathcal{C}}_A(\underset{\sim}{x}'')$ séparable.

16. Sous-Ensembles Attractifs et Répulsifs d'une Surface Limite

Nous introduirons maintenant deux types de sous-ensembles d'une surface limite Σ .

Un sous-ensemble attractif M_{att} de Σ est un ensemble de points interieurs nonréguliers de Σ , tels qu'en chacun d'eux $\overline{\mathcal{C}}_B(\underset{\sim}{x})$ est séparable, mais $\overline{\mathcal{C}}_A(\underset{\sim}{x})$ non séparable.

Un sous-ensemble répulsif M_{rep} de Σ est un ensemble de points interieurs nonréguliers de Σ , tels qu'en chacun d'eux $\overline{\mathcal{C}}_A(\underset{\sim}{x})$ est séparable, mais $\overline{\mathcal{C}}_B(\underset{\sim}{x})$ non séparable.

Ces appellations ont été adoptées comme conséquences des Corollaires suivants

Corollaire 6. Si $\underset{\sim}{x}^*(t')$ et $\underset{\sim}{x}^*(t'')$, $t'' > t'$, sont des points de la trajectoire optimale Γ^* sur la surface limite Σ , et si $\underset{\sim}{x}^*(t')$ appartient à M_{att} , alors $\underset{\sim}{x}^*(t'')$ ne peut pas être un point interieur régulier de Σ ; il ne peut pas non plus appartenir à M_{rep}.

Cependant il n'est pas interdit à une trajectoire optimale issue d'un point interieur régulier de Σ , où d'un point du sous-ensemble répulsif M_{rep} , d'atteindre un point du sous-ensemble attractif M_{att}.

Corollaire 7. Si $\underset{\sim}{x}^*(t')$ et $\underset{\sim}{x}^*(t'')$, $t'' > t'$, sont des points de la trajectoire optimale Γ^* sur la surface limite Σ et si $\underset{\sim}{x}^*(t')$ est un point interieur régulier de Σ, ou s'il appartient à M_{att} alors $\underset{\sim}{x}^*(t'')$ ne peut appartenir à M_{rep} .

Cependant une trajectoire optimale peut sortir du sous-ensemble répulsif M_{rep} .

17. Sous-Ensemble Réguliers et Antiréguliers d'une Surface Limite

Comme nous l'avons noté anterieurement, en un point interieur régulier de Σ, les cones $\overline{\mathcal{C}}_A(\underset{\sim}{x})$ et $\overline{\mathcal{C}}_B(\underset{\sim}{x})$ sont tous deux séparables. Inversement si, en un point interieur de Σ, $\overline{\mathcal{C}}_A(\underset{\sim}{x})$ et $\overline{\mathcal{C}}_B(\underset{\sim}{x})$ sont tous deux séparables, ce point est point régulier de Σ. De plus on peut établir les Corollaires suivants:

Corollaire 8. Si $\underset{\sim}{x}^*(t')$ et $\underset{\sim}{x}^*(t'')$, $t'' > t'$, sont points interieurs réguliers de Σ, et si tous les points $\underset{\sim}{x}^*(t)$, $t' \leq t \leq t''$, sont points interieurs de Σ, alors $\underset{\sim}{x}^*(t)$ est point régulier de Σ.

En effet $\overline{\mathcal{C}}_A(\underset{\sim}{x}^*(t''))$ est séparable, ce qui implique que $\overline{\mathcal{C}}_A(\underset{\sim}{x}^*(t))$ est séparable. D'autre part la séparabilité de $\overline{\mathcal{C}}_B(\underset{\sim}{x}^*(t'))$ implique celle de $\overline{\mathcal{C}}_B(\underset{\sim}{x}^*(t))$. Il s'ensuit que $\underset{\sim}{x}^*(t)$ est point régulier de Σ.

En ce qui concerne les sous-ensembles dont tous les points sont points antiréguliers de Σ, on peut établir aisément que

Corollaire 9. Si $\underset{\sim}{x}^*(t')$ et $\underset{\sim}{x}^*(t'')$, $t'' > t'$, sont des points de la trajectoire optimale Γ^*, sur la surface limite Σ;

(i) si $\underset{\sim}{x}^*(t')$ appartient à un sous-ensemble antirégulier, alors $\underset{\sim}{x}^*(t'')$ ne peut appartenir à un sous-ensemble régulier ni à un sous-ensemble répulsif;

(ii) si $\underset{\sim}{x}^*(t')$ appartient à un sous-ensemble régulier, ou à un sous-ensemble attractif, alors $\underset{\sim}{x}^*(t'')$ ne peut appartenir à un sous-ensemble antirégulier,

De nouveau ces corollaires sont des conséquences directes des Théorèmes I et II .

18. Sous-Ensembles Symmetriques du Cone Local $\mathcal{S}(\underset{\sim}{x})$

Nous appelerons ensemble-symmetrique du cone local $\mathcal{S}(\underset{\sim}{x})$ l'ensemble $I(\underset{\sim}{x})$ des points $\underset{\sim}{x}+\eta$, tels que

$$\underset{\sim}{x} + \eta \in \mathcal{S}(\underset{\sim}{x}) \quad \text{et} \quad \underset{\sim}{x} - \eta \in \mathcal{S}(\underset{\sim}{x})$$

A propos de tels sous-ensembles, nous avons les Lemmes suivants

Lemme 8. Soient $\underset{\sim}{x}' \triangleq \underset{\sim}{x}^*(t')$ et $\underset{\sim}{x}'' \triangleq \underset{\sim}{x}^*(t'')$, $t'' > t'$, des points de la trajectoire optimale Γ^*. Si $\overline{\mathcal{C}}_A(\underset{\sim}{x}'')$ est séparable, alors

$$\underset{\sim}{x}' + \eta' \in I(\underset{\sim}{x}') \Longrightarrow \underset{\sim}{x}'' + \eta'' \in I(\underset{\sim}{x}'')$$

où

$$\eta'' = A(t', t'')\, \eta'$$

Comme $\overline{\mathcal{C}}_A(\underset{\sim}{x}'')$ est séparable par hypothèse , il résulte du Théorème II que $\overline{\mathcal{C}}_A(\underset{\sim}{x}')$ est séparable. Soit $\mathcal{J}(\underset{\sim}{x}'')$ un hyperplan séparant de $\overline{\mathcal{C}}_A(\underset{\sim}{x}'')$. D'après les résultats du paragraphe 14 , $\mathcal{J}(\underset{\sim}{x}'')$ est le transformé (par la transformation linéaire $A(t', t'')$) de l'hyperplan séparant $\mathcal{J}(\underset{\sim}{x}')$ de $\overline{\mathcal{C}}_A(\underset{\sim}{x}')$.

Désignons par $\overline{R}''_A$ et $\overline{R}'_A$ les demi-espaces fermés déterminés par $\mathcal{J}(\underset{\sim}{x}'')$ et $\mathcal{J}(\underset{\sim}{x}')$, respectivement, où

$$\overline{\mathcal{C}}_A(\underset{\sim}{x}') \subseteq \overline{R}'_A \quad \text{et} \quad \overline{\mathcal{C}}_A(\underset{\sim}{x}'') \subseteq \overline{R}''_A$$

Rappelons que $\overline{R}''_A$ est le transformé de $\overline{R}'_A$ (paragraphe 14)

Considérons un vecteur η' en x', tel que

$$x' + \eta' \in I(x')$$

D'après la définition d'un sous-ensemble symmetrique $I(x')$, nous avons

$$x' + \eta' \in \mathcal{I}(x') \qquad \text{et} \qquad x' - \eta' \in \mathcal{I}(x')$$

Ainsi

$$x' + \eta' \in \overline{\mathcal{C}}_A(x')$$

de telle sorte que, d'après le Lemme 6

$$x'' + \eta'' \in \overline{\mathcal{C}}_A(x'')$$

Maintenant si

$$x'' + \eta'' \in \mathring{\mathcal{C}}_A(x'')$$

alors

$$x'' + \eta'' \in R''_A$$

ce qui implique que

$$x'' - \eta'' \in \text{comp } \overline{R}''_A$$

Mais ceci est impossible, puisque

$$x' - \eta' \in \mathcal{I}(x') \Longrightarrow x' - \eta' \in \overline{\mathcal{C}}_A(x')$$

$$\Longrightarrow x' - \eta' \in \overline{R}'_A$$

$$\Longrightarrow x'' - \eta'' \in \overline{R}''_A$$

et par conséquent

$$\underset{\sim}{x}'' + \eta'' \in \mathcal{S}(\underset{\sim}{x}'')$$

De plus, puisque

$$\underset{\sim}{x}' - \eta' \in I(\underset{\sim}{x}')$$

les mêmes arguments sont valables pour $-\eta'$; par conséquent

$$\underset{\sim}{x}'' - \eta'' \in \mathcal{S}(\underset{\sim}{x}'')$$

ce qui établit le Lemme.

De facon analogue on peut établir le Lemme 9

Lemme 9 . Soient $\underset{\sim}{x}' \triangleq \underset{\sim}{x}^*(t')$ et $\underset{\sim}{x}'' \triangleq \underset{\sim}{x}^*(t'')$, $t'' \geq t'$, des points de la trajectoire optimale Γ^*. Si $\mathcal{C}_B(\underset{\sim}{x}')$ est séparable, alors

$$\underset{\sim}{x}'' + \eta'' \in I(\underset{\sim}{x}'') \Longrightarrow \underset{\sim}{x}' + \eta' \in I(\underset{\sim}{x}')$$

où

$$\eta'' = A(t', t'')\, \eta'$$

Ces Lemmes s'appliquent trivialement au cas des vecteurs nuls..

A partir des Lemmes 8 et 9, on peut facilement établir les Théorèmes suivants

Théorème III. Soient $\underset{\sim}{x}' \triangleq \underset{\sim}{x}^*(t')$ et $\underset{\sim}{x}'' \triangleq \underset{\sim}{x}^*(t'')$, $t'' \geq t'$; des points de la trajectoire optimale Γ^* , et soient γ' et γ'' les dimensions de $I(\underset{\sim}{x}')$ et $I(\underset{\sim}{x}'')$, respectivement. Si $\mathcal{C}_A(\underset{\sim}{x}'')$ est séparable , alors $\gamma'' \geq \gamma'$.

Théorème IV. Soient $\underset{\sim}{x}' \triangleq \underset{\sim}{x}^*(t')$ et $\underset{\sim}{x}'' \triangleq \underset{\sim}{x}^*(t'')$, $t'' \geq t'$, des points de la trajectoire optimale Γ^*, et soient γ' et γ''

les dimensions de $I(\underset{\sim}{x}')$ et $I(\underset{\sim}{x}'')$, respectivement. Si $\overline{\mathcal{C}}_B(\underset{\sim}{x}')$ est séparable, alors $\gamma'' \leq \gamma'$.

Théorème V. Si

(i) $I(\underset{\sim}{x}')$ et $I(\underset{\sim}{x}'')$ sont des hyperplans aux points $\underset{\sim}{x}' = \underset{\sim}{x}^*(t')$ et $\underset{\sim}{x}'' = \underset{\sim}{x}^*(t'')$, $t'' \geq t'$, de la trajectoire optimale Γ^*, où $\overline{\mathcal{C}}_A(\underset{\sim}{x}')$ et $\overline{\mathcal{C}}_A(\underset{\sim}{x}'')$, ou $\overline{\mathcal{C}}_B(\underset{\sim}{x}')$ et $\overline{\mathcal{C}}_B(\underset{\sim}{x}'')$, sont séparables; et

(ii) les dimensions de $I(\underset{\sim}{x}')$ et $I(\underset{\sim}{x}'')$ sont égales :

alors $I(\underset{\sim}{x}'')$ est le transformé de $I(\underset{\sim}{x}')$, par la transformation linéaire $A(t', t'')$.

19. Cas Dégénéré

Nous appelerons sous-ensemble dégénéré de Σ un sous-ensemble de Σ en tout point duquel

(i) ou $\overset{\circ}{\mathcal{C}}_B(\underset{\sim}{x}) = \emptyset$

(ii) ou $\overset{\circ}{\mathcal{C}}_A(\underset{\sim}{x}) = \emptyset$

Dans le cas (i) le sous-ensemble dégénéré serà appelé sous-ensemble B-dégénéré, et dans le càs (ii) il sera appelé sous-ensemble A-dégénéré.

Dans ce paragraphe nous examinerons quelques propriétés d'une trajectoire optimale dont une portion (ou un point) appartient à un sous-ensemble dégénéré.

Considérons d'abord le cas d'un sous-ensemble B-dégénéré, c'est-à-dire en tout point duquel

$$\overset{\circ}{\mathscr{C}}_B(\underset{\sim}{x}) = \emptyset \quad , \quad \underset{\sim}{x} \overset{\triangle}{=} \underset{\sim}{x}^*(t)$$

Supposons que

$$\overset{\circ}{\mathscr{C}}_B(\underset{\sim}{x}') = \emptyset \quad , \quad \underset{\sim}{x}' \overset{\triangle}{=} \underset{\sim}{x}^*(t') \quad , \quad t_o \leqslant t \leqslant t' \leqslant t_1$$

et soit un vecteur $\underset{\sim}{\eta}'$ au point $\underset{\sim}{x}'$ tel que

$$\underset{\sim}{x}' + \underset{\sim}{\eta}' \in \overset{\circ}{\mathscr{C}}_B(\underset{\sim}{x}') \quad \text{où} \quad \underset{\sim}{\eta}' \overset{\triangle}{=} \underset{\sim}{\eta}(t')$$

$\underset{\sim}{\eta}'$ est le transformé d'un vecteur $\underset{\sim}{\eta} \overset{\triangle}{=} \underset{\sim}{\eta}(t)$ en $\underset{\sim}{x}$ par la transformation linéaire $A(t, t')$. De plus comme conséquence directe des Lemmes 6 et 7 nous avons

$$\underset{\sim}{x} + \underset{\sim}{\eta} \in \overset{\circ}{\mathscr{C}}_B(\underset{\sim}{x})$$

Mais ceci contredit l'hypothèse. Par conséquent

$$\overset{\circ}{\mathscr{C}}_B(\underset{\sim}{x}') = \emptyset$$

d'où

Théorème VI ;. Si le point $\underset{\sim}{x}^*(t)$ de la trajectoire optimale Γ^* appartient à un sous-ensemble B-dégénéré, alors $\underset{\sim}{x}^*(t')$ appartient à un sous-ensemble B-dégénéré, quel que soit $t' \geqslant t$.

Considérons maintenant le vecteur $\underset{\sim}{\eta}'$ en $\underset{\sim}{x}'$ tel que

$$\underset{\sim}{x}' + \underset{\sim}{\eta}' \in \mathscr{S}(\underset{\sim}{x}')$$

D'après le Lemme 7 il est le transformé du vecteur $\underset{\sim}{\eta} = \underset{\sim}{\eta}(t)$, tel que

$$\underset{\sim}{x} + \underset{\sim}{\eta} \in \overline{\mathscr{C}}_B(\underset{\sim}{x})$$

et puisque nous supposons que $\mathring{\mathcal{C}}_B(\underset{\sim}{x}) = \emptyset$, il s'ensuit que

$$\underset{\sim}{x} + \underset{\sim}{\eta} \in \mathcal{S}(\underset{\sim}{x})$$

d'où

Théorème VII . Le long d'une portion de trajectoire optimale Γ^* qui appartient à un sous-ensemble B-dégénéré

$$\underset{\sim}{x}' + \underset{\sim}{\eta}' \in \mathcal{S}(\underset{\sim}{x}') \Longrightarrow \underset{\sim}{x} + \underset{\sim}{\eta} \in \mathcal{S}(\underset{\sim}{x})$$

où $\underset{\sim}{\eta}' = \underset{\sim}{\eta}(t')$, $\underset{\sim}{\eta} = \underset{\sim}{\eta}(t)$, $\underset{\sim}{\eta}' = A(t,t')\underset{\sim}{\eta}$ $t_o \leqslant t \leqslant t' \leqslant t_1$

$$\underset{\sim}{x} = \underset{\sim}{x}^*(t) \ , \ \underset{\sim}{x}' = \underset{\sim}{x}^*(t')$$

De plus , le long d'une portion de trajectoire optimale qui appartient à un sous-enseble B-dégénéré, le Lemme 5 admet le Corollaire suivant

Corollaire 10. $\underset{\sim}{x} - \underset{\sim}{f}(\underset{\sim}{x}, \underset{\sim}{u}) \in \mathcal{S}(\underset{\sim}{x})$, $\underset{\sim}{x} \overset{\Delta}{=} \underset{\sim}{x}^*(t)$

$$\forall \underset{\sim}{u} \in \mathcal{U}$$

En invoquant des arguments analogues, en ce qui concerne les sous-ensembles A-dégénérés, on aboutit aux conclusions suivantes

Théorème VIII. Si le point $\underset{\sim}{x}^*(t)$ de la trajectoire optimale Γ^* appartient à un sous-ensemble A-dégénéré, alors $\underset{\sim}{x}^*(t'')$ appartient à un sous-ensemble A-dégénéré, quel que soit t'' , $t_o \leqslant t'' \leqslant t \leqslant t_1$.

Théorème IX . Le long d'une portion de trajectoire optimale Γ^* qui appartient à un sous-ensemble A-dégénéré

$$\underset{\sim}{x}'' + \underset{\sim}{\eta}'' \in \mathcal{S}(\underset{\sim}{x}'') \Longrightarrow \underset{\sim}{x} + \underset{\sim}{\eta} \in \mathcal{S}(\underset{\sim}{x})$$

<u>où</u> $\eta'' = \eta(t'')$, $\eta = \eta(t)$, $\eta = A(t'', t)\, \eta''$, $t_o \leqslant t'' \leqslant t \leqslant t_1$

$$\underset{\sim}{x} = \underset{\sim}{x}^*(t) , \quad \underset{\sim}{x}'' = \underset{\sim}{x}^*(t'')$$

<u>Corollaire 11.</u> $\underset{\sim}{x} + \underset{\sim}{f}(\underset{\sim}{x}, \underset{\sim}{u}) \in \mathcal{S}(\underset{\sim}{x})$, $\underset{\sim}{x} \triangleq \underset{\sim}{x}^*(t)$

$$\forall \underset{\sim}{u} \in U$$

Finalement , comme consequence du paragraphe 6, on a

Si $\mathring{\mathcal{C}}_B(\underset{\sim}{x}) = \emptyset$ alors $L_- \subset \mathcal{S}$

Si $\mathring{\mathcal{C}}_A(\underset{\sim}{x}) = \emptyset$ alors $L_+ \subset \mathcal{S}$

<u>20. Principe du Maximum (Points interieurs de Σ , réguliers ou non-réguliers)</u>

Considérons à titre d'exemple une trajectoire optimale Γ^* dont aucun des points n'appartient à un sous-ensemble répulsif de Σ . Soient $\underset{\sim}{x}^o \in E^{n+1}$ et $\underset{\sim}{x}' \in E^{n+1}$ son point de départ et son point terminal (c'est-à-dire la cible), respectivement, aux temps t_o et t_1. $\underset{\sim}{x}^o$ peut être un point régulier ou un point non régulier de Σ , en tous cas $\overline{\mathcal{C}}_B(\underset{\sim}{x}^o)$ est séparable.

Soit $\mathcal{J}(\underset{\sim}{x}^o)$ un hyperplan séparant, et $\overline{R}^o_B$ le demi-espace fermé qui contient $\overline{\mathcal{C}}_B(\underset{\sim}{x}^o)$, suivant les notations du paragraphe 9.

D'après le Théorème I, $\overline{\mathcal{C}}_B(\underset{\sim}{x}^*(t))$ est séparable quel que soit t, $t_o \leqslant t \leqslant t_1$, et d'après les remarques (ii) (iii) du paragraphe 14 le transformé de $\mathcal{J}(\underset{\sim}{x}^o)$, par la transformation linéaire $A(t_o, t)$ est un hyperplan séparant de $\overline{\mathcal{C}}_B(\underset{\sim}{x}^*(t))$, et

$$\mathscr{C}_B(\underset{\sim}{x}^*(t)) \subset \overline{R}_B$$

où $\overline{R}_B$ est le transformé de $\overline{R}_B^\circ$.

Soit $\underset{\sim}{n}^o$, $|\underset{\sim}{n}^o| = 1$, le vecteur normal à $\mathfrak{J}(\underset{\sim}{x}^o)$ en $\underset{\sim}{x}^o$ tel que

$$\underset{\sim}{n}^o \in \overline{R}_B^o$$

et soit $\underset{\sim}{\lambda}(t_o) = \lambda^o\,\underset{\sim}{n}^o$, $\lambda^o > 0$

D'après la remarque (ii) du paragraphe 12

$\underset{\sim}{\lambda}(t)$ est normal a $\mathfrak{J}(\underset{\sim}{x}^*(t))$, transformé de $\mathfrak{J}(\underset{\sim}{x}^o)$ et

$$\underset{\sim}{\lambda}(t) \in \overline{R}_B$$

Finalement, d'après le Lemme 5

$$\dot{\underset{\sim}{x}}^*(t) - \underset{\sim}{f}(\underset{\sim}{x}^*(t), \underset{\sim}{u}) \in \mathscr{C}_B(\underset{\sim}{x}^*(t)) \qquad \forall \underset{\sim}{u} \in \mathcal{U}$$

d'où

$$\dot{\underset{\sim}{x}}^*(t) - \underset{\sim}{f}(\underset{\sim}{x}^*(t), \underset{\sim}{u}) \in \overline{R}_B \qquad \forall \underset{\sim}{u} \in \mathcal{U}$$

Il s'ensuit que

(20) $$\underset{\sim}{\lambda}(t)\cdot\underset{\sim}{f}(\underset{\sim}{x}^*(t), \underset{\sim}{u}) \leqslant 0 \quad \forall \underset{\sim}{u} \in \mathcal{U}$$

Une discussion plus complète conduirait aisément à la condition

(21) $$\underset{\sim}{\lambda}(t)\cdot\underset{\sim}{f}(\underset{\sim}{x}^*(t), \underset{\sim}{u}^*(t)) = 0$$

quel que soit t, $t \in [t_o, t_1]$

et à

(22) $$\lambda_o(t) = \text{constante} \leqslant 0$$

Les conditions (20) - (22) résument le Principe du Maximum dans le cadre de nos hypothèses.

Des conclusions analogues pourraient être obtenues aisément dans le cas de trajectoires optimales appartenant à un sous-ensemble répulsif de Σ (à l'exception de la cible qui est nécessairement régulière ou attractive).

21. Principe du Maximum Trivial

De l'etude du cas dégénéré, il résulte que

(i) si $\underset{\sim}{x}^{o}$ appartient à un sous-ensemble B-dégénéré ;

(ii) si $\mathcal{J}(\underset{\sim}{x}^{o}) \subseteq T^{(k)}$

où $T^{(k)}$ est un hyperplan à k dimensions, $k \leqslant n$:

alors il existe un vecteur non nul $\underset{\sim}{\lambda}(t)$, fonction continue de t, solution de l'équation adjointe (13), tel que

$$\lambda_o(t) = 0$$

$$\underset{\sim}{\lambda}(t) \cdot \underset{\sim}{f}(\underset{\sim}{x}^*(t), \underset{\sim}{u}) = 0 \qquad \forall \underset{\sim}{u} \in \mathcal{U}$$

quel que soit t, $t \in [t_o, t_1]$.

Nous nous limiterons ici à ces quelques remarques concernant le Principe du Maximum, car l'objet de cet exposé n'est pas de redémontrer certaines propriétés bien connues.

III

EXEMPLES ILLUSTRANT LA THEORIE [1]

1. Introduction

Dans le but d'illustrer certains des concepts introduits dans cet exposé, nous discuterons brièvement quelques exemples simples de problèmes faisant intervenir une optimisation de temps de transfert. Nous nous bornerons à l'établissement des équations des surfaces Σ , et nous indiquerons les traits les plus saillants de ces surfaces, dans chaque cas. Le Lecteur pourra développer plus complètement cette discussion, en s'appuyant sur l'étude qui précède.

2. Problème du Régulateur à Une-Dimension

Considérons le système dont les équations d'état sont

$$\dot{x}_1 = x_2$$

$$\dot{x}_2 = u$$

pour lequel l'ensemble de commande U est défini par

$$|u| \leqslant 1$$

On se propose de transferrer le système du point initial (x_1^o, x_2^o) au point terminal $(0,0)$, dans le temps minimum.

(1) A. Blaquière et G. Leitmann, On the Geometry of Optimal Processes - Part III, ONR Report No. AM-66-1.

On sait que la commande optimale est du type bang-bang, et qu'elle comporte au plus une commutation. La courbe de commutation - c'est-à-dire le lieu des points de commutation de la commande - divise l'espace des états E^2 en deux régions ouvertes que nous désignerons par I et II (fig. 1).

Le coût minimal $V^*(\underset{\sim}{x}; \underset{\sim}{x}^1)$ - c'est-à-dire le temps de transfert minimal de l'état $\underset{\sim}{x} = (x_1, x_2)$ à l'état $\underset{\sim}{x}^1 = (0,0)$ - est donné par

$$V^*(\underset{\sim}{x}; \underset{\sim}{x}^1) = \begin{cases} x_2 + \sqrt{2x_2^2 + 4x_1} & \text{si } \underset{\sim}{x} \in \text{région I} \\ -x_2 + \sqrt{2x_2^2 - 4x_1} & \text{si } \underset{\sim}{x} \in \text{région II} \\ \pm\, x_2 & \text{si } \underset{\sim}{x} \in \text{courbe de commutation} \end{cases}$$

Les équations de S et Σ sont

$$S : \quad V^*(\underset{\sim}{x}; \underset{\sim}{x}^1) = C$$

$$\Sigma : \quad x_0 + V^*(\underset{\sim}{x}; \underset{\sim}{x}^1) = C$$

Ces surfaces sont représentées sur les figures 1 et 2, respectivement.

Une surface isocoût optimale S est formée de deux arcs de parabole, S_I et S_{II}. L'arc de parabole de la région I est tangent à la courbe de commutation en A, et l'arc de la région II est tangent à la courbe de commutation en B.

Une surface limite Σ possède une arête ABC dont la projection dans E^2 appartient à la courbe de commutation.

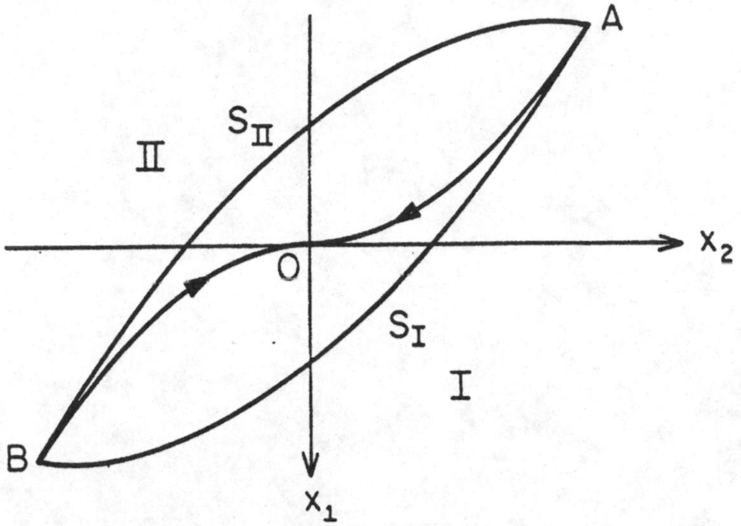

FIG. 1 S-SURFACE

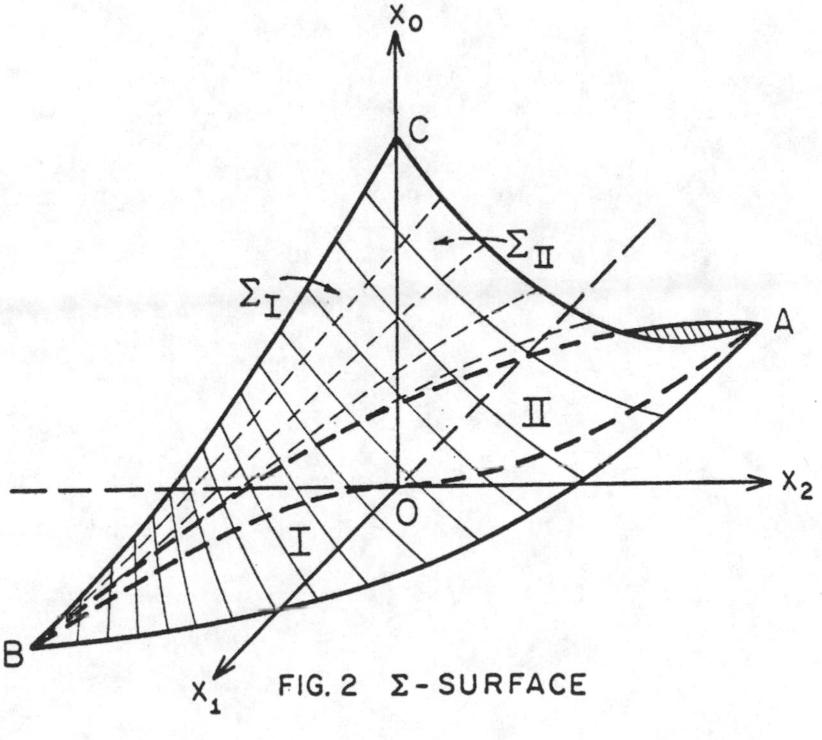

FIG. 2 Σ-SURFACE

L'arête ABC est un sous-ensemble attractif de Σ. Tous les autres points de Σ sont fortement réguliers.

3. Problème du Rocket de Puissance Limitée

Le comportement d'un rocket de puissance limitée en vol rectiligne est décrit par

$$\dot{x}_1 = u$$
$$\dot{x}_2 = u^2$$

L'ensemble de commande $\mathcal{U}$ est défini par

$$|u| \leqslant 1$$

On se propose de transferrer le système du point initial (x^o_1, x^o_2) au point terminal (x^1_1, x^1_2), dans le temps minimum.

La région E^* est le demi-plan ouvert défini par $x_2 < x^1_2$. De plus, cette région est divisée en trois sous-régions I, II, III, par deux demi-droites de pentes 1 et -1, respectivement, (Fig. 3). Si le point de départ (x^o_1, x^o_2) appartient à la région I, toute loi de commande du type bang-bang est optimale (pourvu qu'elle assure le transfert au point terminal $\underset{\sim}{x}^1$). Si le point de départ appartient à la région II ou à la région III, la commande optimale est constante et unique.

Le coût minimum du transfert de $\underset{\sim}{x}$ à $\underset{\sim}{x}^1$ est donné par

$$V^*(\mathfrak{X};\mathfrak{X}^1) = \begin{cases} x_2^1 - x_2 & \text{si } \mathfrak{X} \in \text{région I} \\ \dfrac{(x_1^1 - x_1)^2}{x_2^1 - x_2} & \text{si } \mathfrak{X} \in \text{région II, III} \end{cases}$$

Les frontières des régions I et II, et I et III peuvent être considérees comme appartenant à l'une ou l'autre de ces régions.

Il est facile de voir qu'une surface $\mathcal{S}$ est formée d'un segment de droite $\mathcal{S}_I$, parallèle à l'axe x_1, et d'un arc de parabole, $\mathcal{S}_{II} - \mathcal{S}_{III}$, qui admet l'axe x_2 pour axe de symmétrie, (fig. 3). Une surface Σ est formée d'une portion plane triangulaire, Σ_I, inclinée à 45° par rapport au plan des états, et d'une portion de cone parabolique, $\Sigma_{II} - \Sigma_{III}$ (fig. 4).

Une surface Σ possède deux arêtes, AC et BC, qui sont toutes deux des sous-ensembles attractifs de Σ. Tous les autres points de Σ sont fortement réguliers.

4. Un Problème de Navigation

Les équations d'état d'un véhicule, dont la vitesse de déplacement relatif par rapport à un courant constant est constante en grandeur mais non en direction; sont

$$\dot{x}_1 = s + u_1$$
$$\dot{x}_2 = u_2$$

où s = constante, et $\mathcal{U}$ est défini par

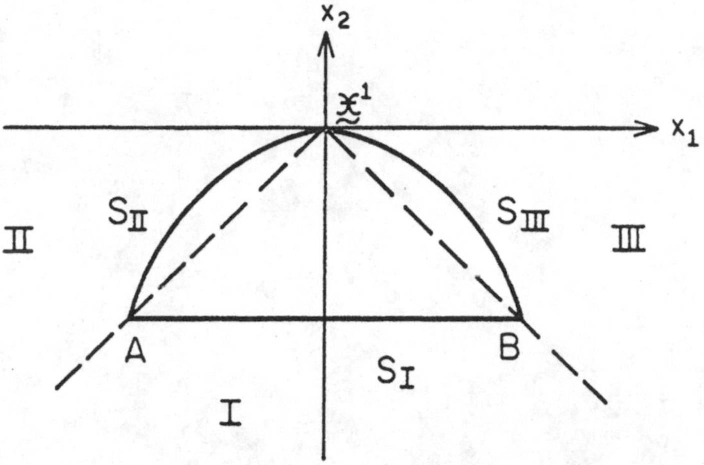

FIG. 3 S-SURFACE

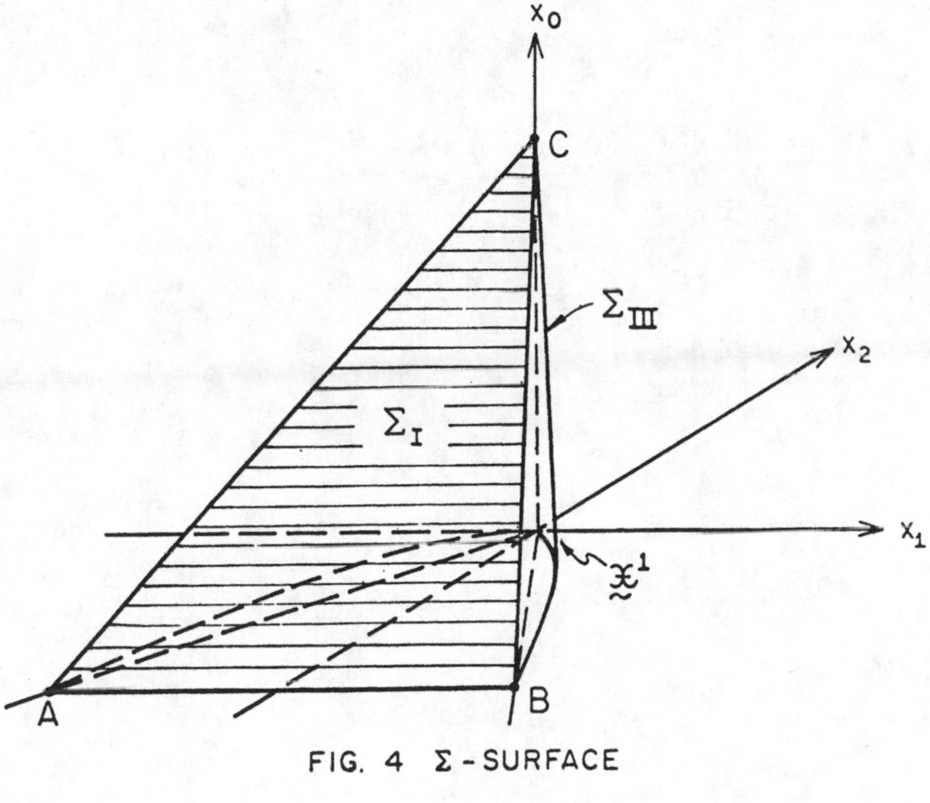

FIG. 4 Σ-SURFACE

$$u_1^2 + u_2^2 = 1$$

On se propose de transferrer le système du point initial (x_1^o, x_2^o) au point terminal (x_1^1, x_2^1), dans le temps minimum.

Ici nous distinguerons trois cas :

(i) $s < 1$; (ii) $s = 1$; et (iii) $s > 1$.

Dans chaque cas la commande optimale est telle que

$$u_1^*(t) \equiv \text{constante}$$

$$u_2^*(t) \equiv \text{constante}$$

Cependant, en ce qui concerne la région E^* et le coût minimum nous devons considérer ces différents cas séparément.

(i) $s < 1$. On a alors $E^* = E^2$, et

$$V^*(\underset{\sim}{x}; \underset{\sim}{x}^1) = \frac{-s\,\Delta x_1 + \sqrt{(\Delta x_1)^2 + (1-s^2)(\Delta x_2)^2}}{1-s^2}$$

où

$$\Delta x_1 \triangleq x_1^1 - x_1$$

$$\Delta x_2 \triangleq x_2^1 - x_2$$

Une surface S est un cercle qui entoure l'état terminal $\underset{\sim}{x}^1$; (fig. 5) Une surface Σ est un cone circulaire, représenté sur la figure 6. Σ est fortement régulière en tout point distinct de la cible.

(ii) $s=1$. Dans ce cas E^* est le demi-plan ouvert $x_1 < x_1^1$, et

$$V^*(\underset{\sim}{x} ; \underset{\sim}{x}^1) = \frac{(\Delta x_1)^2 + (\Delta x_2)^2}{2\,\Delta x_1}$$

Une surface S est un cercle dont le point terminal $\underset{\sim}{x}^1$ se trouve exclu ; $\underset{\sim}{x}^1$ appartient à la frontière de E^*, comme l'indique la figure 7. Une surface Σ est un cone circulaire dont la génératrice parallèle à l'axe x_o est exclue (Fig. 8). De nouveau Σ est fortement régulière en tout point.

(iii) $s > 1$ Ici E^* est une région triangulaire fermée, représentée sur la figure 9. Le coût minimum est donné par

$$V^*(\underset{\sim}{x} ; \underset{\sim}{x}^1) = \frac{s\Delta x_1 + \sqrt{(\Delta x_1)^2 - (s^2 - 1)(\Delta x_2)^2}}{s^2 - 1}$$

Une surface S est un arc de cercle qui est tangent à la frontière de E^*, (Fig. 9). Une surface Σ est une portion de cone circulaire, représentée sur la figure 10. Σ possède des points frontières le long des génératrices AC et BC. Tout point interieur de Σ est fortement régulier.

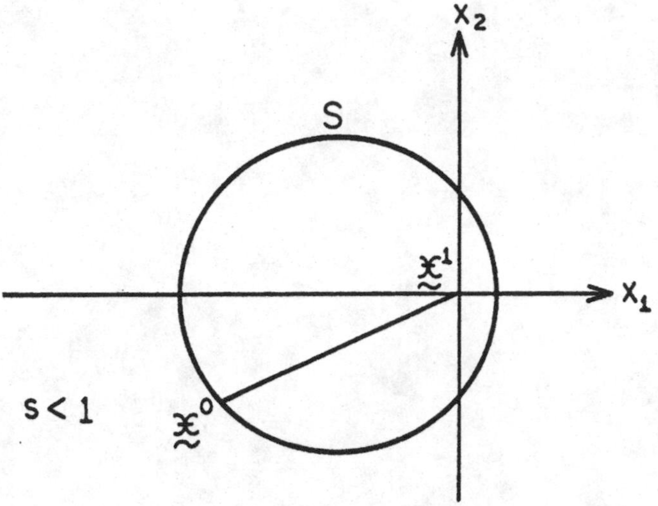

FIG. 5 S - SURFACE

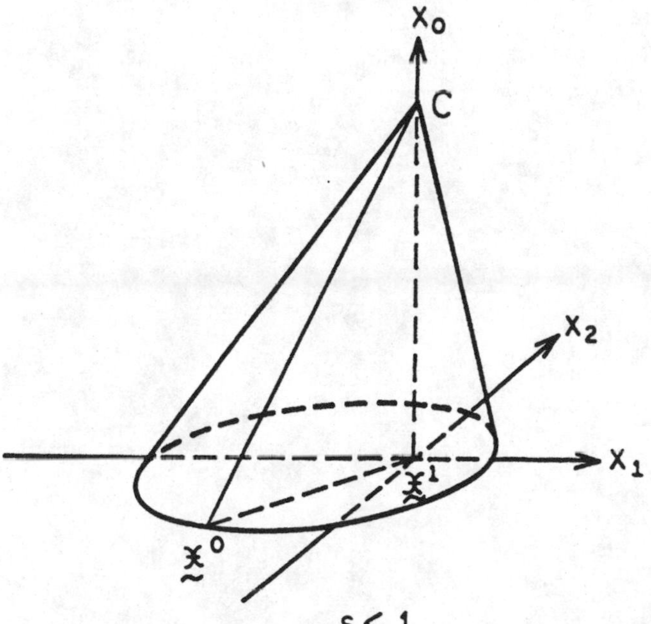

FIG. 6 Σ-SURFACE

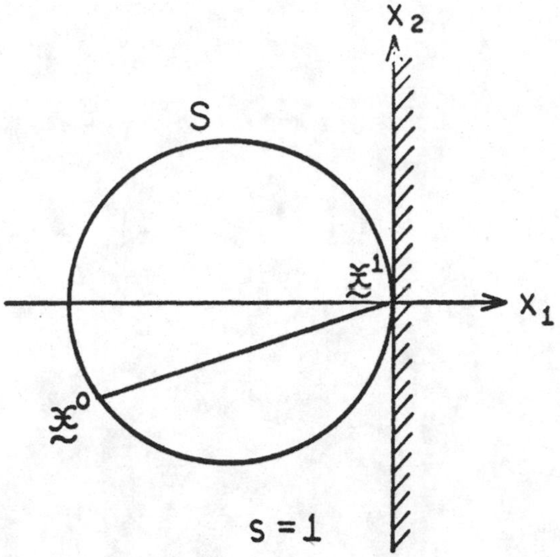

FIG. 7 S-SURFACE

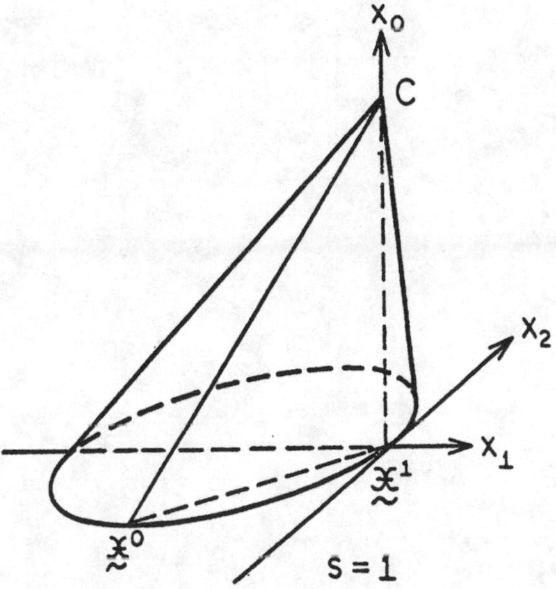

FIG. 8 Σ-SURFACE

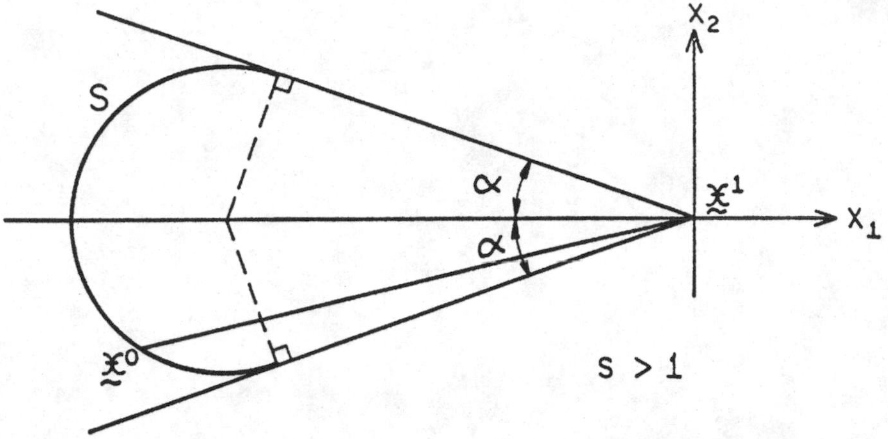

FIG. 9 S - SURFACE

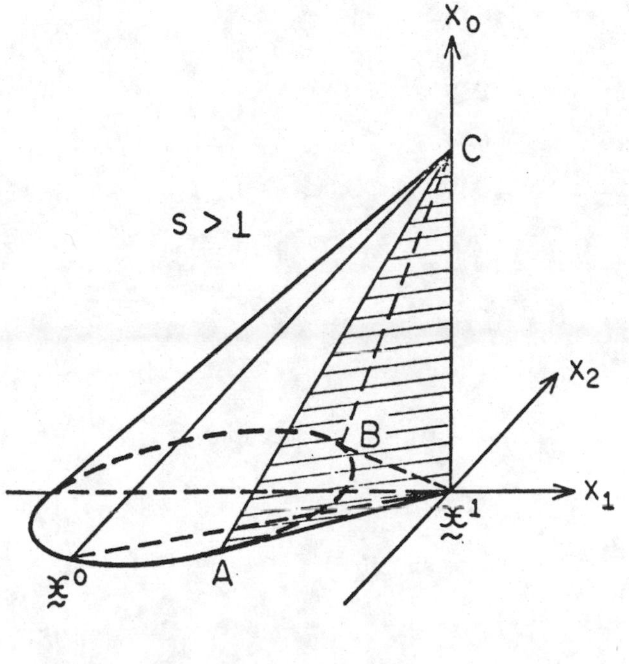

FIG. 10 Σ - SURFACE

CENTRO INTERNAZIONALE MATEMATICO ESTIVO
(C.I.M.E.)

Ch. CASTAING

QUELQUES PROBLEMES DE MESURABILITE LIES A LA THEORIE DES COMMANDES.

Corso tenuto a Bressanone dal 10 al 18 giugno
1966

QUELQUES PROBLEMES DE MESURABILITE LIES A LA THEORIE DES COMMANDES

par

Ch. CASTAING

DEFINITION 1. - Une multi-application (i.e. application multivoque) Γ d'une espace topologique T dans un espace topologique E est dite semi-continue inférieurement si l'ensemble

$$\Gamma^{-}\Omega = \{t \in T / \Gamma(t) \cap \Omega \neq \emptyset\}$$

est ouvert dans T pour tout ouvert Ω dans E.

DEFINITION 2. - Une multi-application Γ d'un espace topologique T dans un espace topologique séparé E est dite semi-continue supérieurement si $\Gamma(t)$ est compact pour tout $t \in T$ et si l'ensemble

$$\Gamma^{-} F = \{t \in T / \Gamma(t) \cap F \neq \emptyset\}$$

est fermé dans T pour tout fermé F dans E.

Une multi-application Γ est dite continue si elle est semi-continue inférieurement et semi-continue supérieurement.

DEFINITION 3. - Soit T un espace localement compact muni d'une mesure de Radon μ. Une multi-application Γ de T dans un espace topologique E sera dite μ-mesurable si l'ensemble

$$\Gamma^{-} F = \{t \in T / \Gamma(t) \cap F \neq \emptyset\}$$

est μ-mesurable pour tout fermé F dans E.

Toute multi-application semi-continue supérieurement est μ-mesurable pour toute mesure μ sur T.

Pour toute partie M de T, on désigne par $\Gamma|_M$ la restriction de Γ à M.

THEOREME 1. - Soient T un espace localement compact, μ une mesure (de Radon) sur T et Γ une multi-application de T à valeurs dans l'ensemble des compacts non vides d'un espace métrisable séparable E. Les conditions suivantes sont équivalentes

a) Γ est μ-mesurable.

b) Pour toute partie compacte $X \subset T$ et tout $\varepsilon > 0$, il existe une partie compacte $K \subset X$ telle que $|\mu|(X-K) \leq \varepsilon$ et telle que $\Gamma|_K$ soit continue (*)

THEOREME 2. - Soient T un espace localement compact, μ une mesure sur T et Γ une multi-application μ-mesurable de T à valeurs dans l'ensemble des compacts non vides d'un espace métrisable séparable E. Alors, il existe une application μ-mesurable σ de T dans E telle que $\sigma(t) \in \Gamma(t)$, $\forall t \in T$.

Démonstration. - Soit (u_n) une suite partout dense dans E.

(*) La démonstration de ce théorème fondamental est omise et sera publiée prochainement dans un article promis au Bulletin de la Société mathématique de France.

Pour chaque $t \in T$, on pose

$$\Gamma_1(t) = \left\{ u \in \Gamma(t) \;/\; d(u, u_1) = \inf_{v \in \Gamma(t)} d(v, u_1) \right\}$$

et pour $n > 1$

$$\Gamma_n(t) = \left\{ u \in \Gamma_{n-1}(t) \;/\; d(u, u_n) = \inf_{v \in \Gamma_{n-1}(t)} d(v, u_n) \right\}$$

Grâce au théorème 1, on démontre que chacune des multi-applications Γ_n est μ-mesurable. Il en est de même de la multi-application $t \dashrightarrow \bigcap_{n=1}^{\infty} \Gamma_n(t)$. Il suffit de vérifier que $\bigcap_{n=1}^{\infty} \Gamma_n(t)$ est réduit à un point $\sigma(t)$ pour chaque $t \in T$. Si $\sigma'(t) \in \bigcap_{n=1}^{\infty} \Gamma_n(t)$ on a

$$d(\sigma'(t), u_n) = d(\sigma(t), u_n), \forall_n . \text{ Donc } \sigma'(t) = \sigma(t), \; \forall t \in T .$$

COROLLAIRE . - Soient h une application continue de $T \times E$ dans un espace topologique séparé G et g une application μ-mesurable de T dans G telle que $g(t) \in h(t, \Gamma(t))$, $\forall t \in T$. Alors il existe une application mesurable s de T dans E telle que $s(t) \in \Gamma(t)$ et $g(t) = h(t, s(t))$, $\forall t \in T$.

Démonstration. - Pour toute partie compacte $X \subset T$ et tout $\varepsilon > 0$, il existe une partie compacte $K \subset T$ telle que $|\mu|(X-K) \leq \varepsilon$ et telle que $g|_K$ et $\Gamma|_K$ soient continues. Il en résulte que la restriction à K de la multi-application Σ de T dans E définie par :

$$\Sigma(t) = \left\{ u \in E / h(t, u) = g(t) \right\}$$

est de graphe fermé. Par suite, la restriction à K de la multi-application S de T dans E définie par :

$$S(t) = \{u \in \Gamma(t) / h(t, u) = g(t)\}$$

est semi-continue supérieurement. On en conclut qu'il existe dans X une partie négligeable N et une partition dénombrable de X-N formée de compacts K_n telle que $S|_{K_n}$ soit semi-continue supérieurement pour tout n. Si F est un ensemble fermé dans E, $\bar{S} F \cap X$ est réunion de $\bar{S} F \cap N$ et des compacts $\bar{S} F \cap K_n$, donc S est μ-mesurable sur T entier. En vertu du théorème 2, il existe une application μ-mesurable s de T dans E telle que $s(t) \in S(t)$, $\forall t \in T$ et ceci équivaut à $s(t) \in \Gamma(t)$ et $g(t) = h(t, s(t))$, $\forall t \in T$; ce qui termine la démonstration.

THEOREME 3. - Soient T un espace localement compact, μ une mesure sur T, E un espace localement compact à base dénombrable, G un espace métrisable séparable et h une application de T x E dans G telle que 1°) pour tout $t \in T$ $u \longrightarrow h(t, u)$ est continue sur E 2°) pour tout $u \in E$, $t \longrightarrow h(t, u)$ est μ-mesurable.

Alors, pour toute partie compacte $X \subset T$ et tout $\varepsilon > 0$, il existe une partie compacte $K \subset X$ telle que $|\mu|(X-K) \leq \varepsilon$ et que la restriction de h à KxE soit continue.

Du théorème précédent on deduit le résultat suivant qui est une variante du corollaire 1 du théorème 2.

COROLLAIRE 2. - Soient T un espace localement compact, μ une mesure sur T. E un espace localement compact à base

dénombrable, Γ une multi-application μ-mesurable de T à valeurs dans l'ensemble des compacts non vides de E, h une application de $T \times E$ dans un espace métrisable séparable G telle que 1°) pour tout $t \in T$, $u \longrightarrow h(t, u)$ soit continue sur E. 2°) pour tout $u \in E$, $t \longrightarrow h(t, u)$ soit μ-mesurable, et enfin g une application μ-mesurable de T dans G telle que $g(t) \in h(t, \Gamma(t))$, $\forall t \in T$.
Alors, il existe une application mesurable s de T dans E telle que $s(t) \in \Gamma(t)$ et $g(t) = h(t, s(t))$, $\forall t \in T$.

THEOREME 4. - Soient T un espace localement compact, μ une mesure sur T, E un espace polonais et Γ une multi-application μ-mesurable de T à valeurs dans l'ensemble des fermés non vides de E. Alors, il existe une application mesurable σ de T dans E telle que $\sigma(t) \in \Gamma(t)$, $\forall t \in T$.

Démonstration. - Soit (C_n, p_n, φ_n) un criblage de E. On munit chacun des ensembles C_o et $(p_n^{-1}(c)\ ;\ c \in C_n)$ d'une relation d'ordre total telle que l'ensemble des éléments inférieurs à un élément donné soit fini. Soit $c \in C_o$ et soit T_c l'ensemble ainsi défini

$$t \in T_c \Longleftrightarrow c \text{ est le premier élément de } C_o \text{ tel que } \Gamma(t) \cap \varphi_o(c) \neq \emptyset .$$

Les ensembles T_c sont μ-mesurables et forment une partition dénombrable de T.
Supposons déjà défini T_c pour $c \in C_n$. Soit $c' \in p_n^{-1}(c)$ et $T_{c'}$ l'ensemble ainsi défini

$$t \in T_{c'} \Longleftrightarrow \begin{cases} t \in T_c \\ c' \text{ est le premier élément de } p_n^{-1}(c) \text{ tel que} \\ \Gamma(t) \cap \varphi_{n+1}(c') \neq \emptyset \end{cases}$$

Les ensembles $T_{c'}$, sont μ-mesurables et $T_c = \bigcup_{c' \in p_n^{-1}(c)} T_{c'}$.

Pour chaque n, on pose $\sigma_n(t) = u_c \in \varphi_n(c)$ si $t \in T_c$. Chacune des applications σ_n est μ-mesurable et la suite $(\sigma_n(t))$ est de Cauchy. L'application $\sigma : t \longrightarrow \lim_n \sigma_n(t)$ est μ-mesurable (th. EGOROV) et $\sigma(t) \in \Gamma(t)$, $\forall t \in T$ car

$$\lim_n d(\sigma_n(t), \Gamma(t)) = d(\sigma(t), \Gamma(t)) = 0, \ \forall t \in T .$$

COROLLAIRE.- <u>Soient</u> T <u>un espace localement compact tel que toute partie compacte de</u> T <u>soit métrisable,</u> μ <u>une mesure sur</u> T, Γ <u>une multi-application de</u> T <u>à valeurs dans l'ensemble des fermés non vides d'un espace polonais</u> E <u>telle que la restriction de</u> Γ <u>à toute partie compacte de</u> T <u>soit de graphe sousli-nien. Soient</u> h <u>une application continue de</u> $T \times E$ <u>dans un espace topologique séparé</u> G <u>et</u> g <u>une application</u> μ<u>-mesurable de</u> T <u>dans</u> G <u>telle que</u> $g(t) \in h(t, \Gamma(t))$, $\forall t \in T$.
<u>Alors il existe une application mesurable</u> s <u>de</u> T <u>dans</u> E <u>telle que</u> $s(t) \in \Gamma(t)$ <u>et</u> $g(t) = h(t, s(t))$, $\forall t \in T$.

Remarques. - 1) Le résultat précédent est encore vrai dans le cas où E est un espace localement compact à base dénombrable, G un espace métrisable séparable et h une application de $T \times E$

dans G telle que 1°) pour tout $t \in T$, $u \longrightarrow h(t,u)$ soit continue sur E. 2°) pour tout $u \in E$, $(t \longrightarrow h(t,u)$ soit μ-mesurable.

2) Signalons enfin que les applications des résultats précédents à la théorie des commandes sont nombreuses et importantes.

CENTRO INTERNAZIONALE MATEMATICO ESTIVO

(C. I. M. E.)

L. CESARI

"EXISTENCE THEOREMS FOR LAGRANGE AND PONTRYAGIN PROBLEMS OF THE CALCULUS OF VARIATIONS AND OPTIMAL CONTROL. MORE DIMENSIONAL EXTENSIONS IN SOBOLEV SPACES."

Corso tenuto a Bressanone dal 10 al 18 giugno

1966

EXISTENCE THEOREMS FOR LAGRANGE AND PONTRYAGIN PROBLEMS OF THE CALCULUS OF VARIATIONS AND OPTIMAL CONTROL. MORE DIMENSIONAL EXTENSIONS IN SOBOLEV SPACES (*)

by

Lamberto Cesari

(Univ. of Michigan, Ann Arbor)

Lecture 1. Usual and generalized solutions in optimal control and the calculus of variations.

1. Usual solutions.

Let A be a closed subset of the tx-space $E_1 \times E_n$, $t \in E_1$, $x = (x^1, \dots, x^n) \in E_n$, and for each $(t, x) \in A$, let $U(t, x)$ be a closed subset of the u-space E_m, $u = (u^1, \dots, u^m)$. We do not exclude that A coincides with the whole tx-space and that U coincides with the whole u-space. Let M denote the set of all (t, x, u) with $(t, x) \in A$, $u \in U(t, x)$. Let $\bar{f}(t, x, u) = (f_0, f) = (f_0, f_1, \dots, f_n)$ be a continuous vector function from M into E_{n+1}. Let B be a closed subset of points (t_1, x_1, t_2, x_2) of E_{2n+2}, $x_1 = (x_1^1, \dots, x_1^n)$, $x_2 = (x_2^1, \dots, x_2^n)$. We shall consider the class of all pairs $x(t)$, $u(t)$, $t_1 \leq t \leq t_2$, of vector functions $x(t)$, $u(t)$ satisfying the following conditions:

(a) $x(t)$ is absolutely continuous (AC) in $[t_1, t_2]$;

(b) $u(t)$ is measurable in $[t_1, t_2]$;

(c) $(t, x(t)) \in A$ for every $t \in [t_1, t_2]$;

(d) $u(t) \in U(t, x(t))$ almost everywhere (a.e.) in $[t_1, t_2]$;

(e) $f_0(t, x(t), u(t))$ is L-integrable in $[t_1, t_2]$;

(f) $dx/dt = f(t, x(t), u(t))$ a.e. in $[t_1, t_2]$;

(g) $(t_1, x(t_1), t_2, x(t_2)) \in B$.

By (f) we mean that the n ordinary differential equations

(*) Research partially supported by AFOSE Grant 942-65 at the University of Michigan

$$(1) \qquad dx^i/dt = f_i(t, x(t), u(t)) , \qquad i = 1, \ldots, n ,$$

are satisfied a.e. in $[t_1, t_2]$. Since x(t) in AC, that is, each component $x^i(t)$ of x(t) is AC, we conclude that all $f_i(t, x(t), u(t))$, $i = 1, \ldots, n$ are L-integrable in $[t_1, t_2]$ as f_0.

A pair x(t), u(t) satisfying (a b c d e f g) is said to be admissible and for such a pair x(t) is called a trajectory and u(t), a strategy control, or steering function. As usual, U(t, x) is said to be the control space at the time t and space point x. The functional

$$(2) \qquad I[x, u] = \int_{t_1}^{t_2} f_0(t, x(t), u(t))dt$$

is called the cost functional, and we seek the minimum of $I[x, u]$ in the total class Ω of admissible pairs x(t), u(t), or in some well defined subclass of Ω.

In the particular case where U(t, x) is a compact subset of E_m for every $(t, x) \in A$, the problem of the minimum of $I[x, u]$ is called a <u>Pontryagin problem</u> of optimal control theory. The general case above, where U(t, x) is a closed but not necessarily compact subset of E_m for every $(t, x) \in A$ will be denoted a <u>Lagrange problem with unilateral constraints</u>. The classical Lagrange problem corresponds essentially to the case where $U = E_m$ is the whole u-space, with the side conditions being here differential equations in normal form.

In the particular case in which $f_0 = 1$, then $I[x, u] = t_2 - t_1$, and the problem of minimization under consideration is then called a problem of <u>minimum transfer time</u> (from the state $x(t_1)$ to the state $x(t_2)$.

There is another particular case of the Lagrange problem which shall be taken into consideration, namely $m = n$, $U = E_m$, and the vector function $f(t, x, u)$ given by $f(t, x, u) = u$, or $f_i(t, x, u) = u^i$, $i = 1, \dots, m = n$, and hence $\tilde{f}(t, x, u) = (f_0, u)$. Then the differential system (1) reduces to $dx^i/dt = u^i$, $i = 1, \dots, n$, and the cost functional becomes

$$(3) \qquad I[x] = \int_{t_1}^{t_2} f_0(t, x(t), x'(t))\, dt .$$

This problem is called a free problem.

2. Generalized solutions. Often a given problem has no optimal solution, but the mathematical problem and the corresponding concept of solution can be modified in such a way that an optimal solution exists and yet neither the system of trajectories, nor the corresponding values of the cost functional are essentially modified. The modified (or generalized) problem and its solutions are of interest in themselves, and have relevant physical interpretations. Essentially, we consider a finite system of distinct strategies which are thought of as being used at the same time according to some probability distribution.

Instead of considering the usual cost functional, differential equations and constraints

$$I[x, u] = \int_{t_1}^{t_2} f_0(t, x(t), u(t))dt,$$

$$(4) \qquad dx/dt = f(t, x(t), u(t)), \qquad f = (f_1, \dots, f_n),$$

$$(t, x(t)) \in A, \qquad u(t) \in U(t, x(t)) ,$$

we consider a new cost functional differential equations and constraints

$$J(x,p,v) = \int_{t_1}^{t_2} g_0(t, x(t), p(t), v(t))dt ,$$

(5) $$dx/dt = g(t, x(t), p(t), v(t)) , \qquad g = (g_1, \ldots, g_n),$$

$$(t, x(t)) \in A, \qquad v(t) \in V(t, x(t)), \qquad p(t) \in \Gamma \ .$$

Precisely, $v(t) = (u^{(1)}, \ldots, u^{(\gamma)})$ represents a finite system of $\gamma \geq n+1$ ordinary strategies $u^{(1)}, \ldots, u^{(y)}$, each $u^{(j)}$ having its values in $U(t, x(t)) \subset E_m$. Thus, we think of $v = (u^{(1)}, \ldots, u^{(\gamma)})$ as a vector variable whose n components $u^{(1)}, \ldots, u^{(\gamma)})$ are themselves vectors u with values in $U(t,x)$. In other words

$$v = (u^{(1)}, \ldots, u^{(\gamma)}), \ u^{(j)} \in U(t,x), \ j = 1, \ldots, \gamma ,$$

or

(6) $$v \in V(t, x) = [U(t, x)]^{\gamma} = Ux\ldots xU \subset E_{m\gamma} ,$$

where the last term is the product space of U by itself taken γ times, and thus V is a subset of the Euclidean space $E_{m\gamma}$. In (5) $p = (p_1, \ldots, p_\gamma)$ represents a probability distribution. Hence, p is an element of the simplex Γ of the Euclidean space E_γ defined by $p_j \geq 0, p_1 + \ldots + p_\gamma = 1$. Finally, in (5) the new control variable is (p, v), with values $(p, v) \in \Gamma \times V(t,x) \subset E_{\gamma + m\gamma}$. In (5) $g = (g_1, \ldots, g_n)$, and all $g_0, g_1, \ldots, g_n$ are defined by

(7) $$g_i(t, x, p, v) = \sum_{j=1}^{\gamma} p_j(t, x, u^{(j)}) , \quad i = 0, \ldots, n \ .$$

As usual we shall require that the functions $p(t)$, $v(t)$, $t_1 \leq t \leq t_2$, are measurable and that $x(t)$, $t_1 \leq t \leq t_2$. is absolutely continuous. As in no . 1 we shall require that $x(t)$ satisfies boundary conditions of the type $(t_1, x(t_1), t_2, x(t_2)) \in B \subset E_{2m+2}$, where B is a given closed subset of E_{2n+2}. As in no . 1 we require $g_0(t, x(t), p(t), v(t))$ to be L-integrable in $[t_1, t_2]$.

We shall say that $[p(t), v(t)]$ is a generalized strategy, that $p(t) = (p_1, \dots, p_\gamma)$ is a probability distribution, and that $v(t) = (u^{(1)}, \dots, u^{(\gamma)})$ is a finite system of (ordinary) strategies. We shall say that $x(t)$ is a generalized trajectory.

It is important to note that any (ordinary) strategy $u(t)$ and corresponding (ordinary) trajectory $x(t)$ (thus, satisfying (1)) can be interpreted as a generalized strategy and generalized trajectory, by taking $v(t) = (u^{(j)}(t), j=1,\dots,\gamma)$ and $p(t) = (p_j(t), j=1,\dots,\gamma)$ defined by $u^{(j)}(t) = u(t)$, $p_j(t) = 1/\gamma$, $j=1,\dots,\gamma$.
Then relations (5) reduce to relations (1).

Instead of the usual set M we shall now consider the set $N \subset E_{1+n+\gamma+m\gamma}$ of all (t, x, p, v) with $(t,x) \in A$, $p \in \Gamma$, $v \in V(t,x)$. As usual, we shall assume that A is a closed subset of $E_1 \times E_n$, and that $\tilde{f} = (f_0, f_1, \dots, f_n)$ is a continuous vector function from M into E_{n+1}.

Under hypotheses which are often satisfied, any generalized trajectory can be approached as much as we want by means of usual solutions, and correspondingly the value of the cost $J[x, p, v]$ can be approached as much as we want by the value of the usual cost $I[x, u]$. In this sense we shall understand that the usual solutions and the corresponding values of the cost functional are not essentially modified by the introduction of generalized solutions.

3. The distance function ρ. If we denote by X the space of all continuous vector functions $x(t) = (x^1, \dots, x^n)$, $a \leq t \leq b$, from arbitrary finite intervals $[a, b]$ to E_n it is convenient to define a distance function $\rho(x, y)$ for elements $x(t)$, $a \leq t \leq b$, and $y(t)$, $c \leq t \leq d$, of X, so as to make X a metric space. For this purpose we extend

$x(t)$ in all $(-\infty, +\infty)$ by defining it equal to $x(a)$ for $t \leq a$ and equal to $x(b)$ for $t \geq b$, and, analogously, for $y(t)$. We then define

$$\rho(x, y) = |a - c| + |b - d| + \max|x(t) - y(t)| ,$$

where the maximum is taken for all $t, -\infty < t < +\infty$. Then ρ is a distance function and X is a metric space.

Every element $x(t)$ of an admissible pair $[x(t), u(t)]$ is an element of X but, of course, the converse is not true.

A class Ω of admissible pairs is said to be <u>complete</u> provided it satisfies the following property: If $x_k(t)$, $u_k(t)$, $t_{1k} \leq t \leq t_{2k}$, $k = 1, 2, \dots$, and $x(t)$, $u(t)$, $t_1 \leq t \leq t_2$, are all admissible pairs, if $x_k(t) \to x(t)$ as $k \to \infty$ in the ρ-metric, and if all pairs $x_k(t)$, $u_k(t)$, $k = 1, 2, \dots$, belong to Ω, then $x(t)$, $u(t)$ also belongs to Ω. The classes usually taken into consideration in applications are complete. The class of all admissible pairs (satisfying) (a b c d e f g)) is certainly complete.

Lecture 2. Upper semicontinuity of variable sets and generalizations.

4. Upper semicontinuity of variable sets.

Using the same notations of no. 1, we shall denote by $Q(t, x)$ the set

$$Q(t,x) = f(t,x,U(t,x)) = [z \in E_n \mid z = f(t,x,u), u \in U(t,x)] \subset E_n .$$

Here $Q(t, x)$ is the image in E_n of the set $U(t, x)$ in the mapping $U(t, x) \rightarrow E_n$ defined by $z = f(t, x, u)$, $u \in U(t, x)$. If f is continuous, as assumed in no. 1, and if $U(t, x)$ is compact for every $(t, x) \in A$, then also $Q(t, x)$ is compact.

Given any point $(t_0, x_0) \in A$ and $\delta > 0$, we denote by $N_\delta(t_0, x_0)$ the set of all $(t, x) \in A$ at a distance $\leq \delta$ from (t_0, x_0). The set $U(t, x)$ is said to be an upper semicontinuous function of (t, x) in A provided for every $(t_0, x_0) \in A$ there is a $\delta > 0$ such that

$$U(t,x) \subset [U(t_0, x_0)]_\epsilon$$

for every $(t, x) \in N_\delta(t_0, x_0)$, where U_ϵ denotes the closed ϵ-neighborhood of U in E_m.

If $U(t, x)$ is compact for every $(t, x) \in A$, and an upper semicontinuous function of (t, x) in the closed set A, then it is easy to prove that $Q(t, x)$ is also compact for every $(t, x) \in A$ and an upper semicontinuous function of (t, x) in A (See [1 e] for details).

5. In the case in which $U(t, x)$ is only closed for every $(t, x) \in A$, the situation above changes entirely. This can be seen by means of some examples.

First let us take $n = m = 1$, $A = [(t,x) \mid t \geq 0,\ x \in E_1]$, $U = U(t,x) = [u \mid u \geq 0]$, and $f(t,x,u) = (u+1)^{-1}$. Then the set U is a fixed half straight line, hence closed and not compact, while its image $Q = Q(t,x) = [z \mid 0 < z \leq 1]$ is not closed. Here U and Q are constant sets, and then certainly upper semicontinuous functions of (t,x).

As a second example let us take $n=1$, $m=2$, $A = [(t,x) \mid t \geq 0,\ x \in E_1]$, and $U = U(t,x) = [u = (u^1, u^2) \mid u^1 \geq 0,\ 0 \leq u^2 \leq t\, u^1]$. Here U depends on the only , and represents the angle between 0 and $\pi/2$ of the first quadrant of the $u^1 u^2$-plane of vertex the origin, spanned by the positive u^1-axis, and the half-straight line of slope t. Obviously, $U(t,x)$ is not an upper semicontinuous function of (t,x) in A.

Finally, we shall take $n=2$, $m=2$, $A = [(t,x) \mid t \geq 0,\ x \in E_1]$, $U(t,x) = [u = (u^1, u^2) \mid u^1 \geq 0,\ 0 \leq u^2 \leq 1]$, and $f = (f_1, f_2)$ with $f_1 = u^1$, $f_2 = t u^1 u^2$. Obviously U is a constant closed set and thus an upper semicontinuous function of (t,x) in A. On the other hand $Q = Q(t,x) = [z = (z^1, z^2) \mid z^1 \leq 0,\ 0 \leq z^2 \leq t z^1]$ is the same as the set U of the previous example, and thus Q is not an upper semicontinuous function of(t,x) in A .

6. Properties (U) and (Q) of variable sets.

If E denotes any subset of E_n we shall denote by cl E the closure of E and by co E the convex hull of E. Thus clco E denotes the closure of the convex hull of E, or briefly , the closed convex hull of E .

Let $U(t,x)$, $(t,x) \in A$, be a variable subset of E_m. For every $\delta > 0$, let $U(t,x,\delta) = \bigcup U(t',x')$, where the union is taken for all $(t',x') \in N_\delta(t,x)$. We shall say that $U(t,x)$ has <u>property (U)</u> at $(\bar{t},\bar{x}) \in A$ if

$$U(\bar{t},\bar{x}) = \bigcap_{\delta>0} \text{cl}\, U(\bar{t},\bar{x},\delta) = \bigcap_{\delta>0} \text{cl} \bigcup_{(t,x)\in N_\delta(\bar{t},\bar{x})} U(t,x).$$

We shall say that $U(t,x)$, $(t,x)\in A$, has property (U) in A if $U(t,x)$ has property (U) at every point $(t,x)\in A$. If a set $U(t,x)$ has property (U), say at $(\bar{t},\bar{x})$, then obviously $U(\bar{t},\bar{x})$ is closed for it is the intersection of closed sets.

For closed sets, property (U) is a generalization of the concept of upper semicontinuity. Indeed, if $U(t,x)$ is closed for every $(t,x)\in A$ and an upper semicontinuous function of (t,x) in A, then $U(t,x)$ has property (U) in A, as it can be proved directly. On the other hand, the second example of no. 5 shows a closed set $U(t,x)$ which has property (U) at every $(t,x)\in A$, but which is not upper an semicontinuous function of (t,x) in A.

Let $Q(t,x)$, $(t,x)\in A$, be a variable subset of E_n. For every $(t,x)\in A$ and $\delta > 0$, let $Q(t,x,\delta) = \bigcup Q(t',x')$ where the union is taken for all $(t',x')\in N_\delta(t,x)$. We shall say that $Q(t,x)$ has property (Q) at $(\bar{t},\bar{x})\in A$ if

$$Q(\bar{t},\bar{x}) = \bigcap_{\delta>0} \text{cl co}\, Q(\bar{t},\bar{x},\delta) = \bigcap_{\delta=0} \text{cl co} \bigcup_{(t,x)\in N_\delta(\bar{t},\bar{x})} Q(t,x).$$

We shall say that $A(t,x)$ has property (Q) in A if $Q(t,x)$ has property (Q) at every point $(t,x)\in A$. If a set $Q(t,x)$ has property (Q), say at $(\bar{t},\bar{x})$, then obviously $Q(\bar{t},\bar{x})$ is closed and convex, as the intersection of closed and convex sets.

For closed convex sets, property (Q) is a generalization of the concept of upper semicontinuity, Indeed, if $Q(t,x)$ is closed and convex for every $(t,x)\in A$, and an upper semicontinuous function of (t,x) in A, then $Q(t,x)$ has property (Q) in A, as it can be proved directly.

On the other hand, the third example of no. 5 shows a closed convex set $Q(t, x)$ which has property (Q) at every $(t, x) \in A$, but which is not an upper semicontinuous function of (t, x) in A. The following lemma will be needed.

(i) If A is closed and $U(t, x)$ has property (U) in A, then the set M is closed.

Proof. If $(\bar{t}, \bar{x}, \bar{u}) \in \mathrm{cl}\ M$ and $\epsilon > 0$ any number, then there are ∞ many points $(t, x, u) \in M$ with $|t-\bar{t}| < \epsilon, |x-\bar{x}| < \epsilon, |u-\bar{u}| < \epsilon$. Thus, $(\bar{t}, \bar{x}) \in A$ since A is closed, $(t, x) \in N_{2\epsilon}(\bar{t}, \bar{x})$, $u \in U(t, x)$, and finally $\bar{u} \in U(\bar{t}, \bar{x}, 2\epsilon)$ as well as $\bar{u} \in \mathrm{cl}\ U(\bar{t}, \bar{x}, 2\epsilon)$. Thus $\bar{u} \in \bigcap \mathrm{cl}\ U(\bar{t}, \bar{x}, 2\epsilon)$, where the intersection is taken for all $\epsilon > 0$. Hence $\bar{u} \in U(\bar{t}, \bar{x})$ since $U(t, x)$ has property (U) at $(\bar{t}, \bar{x})$. This proves that $(\bar{t}, \bar{x}, \bar{u}) \in M$, that is, M is closed.

For more details on the interrelations between upper semicontinuity, property (U) and property (Q) of variable sets, see [1. e].

Lecture 3. Closure Theorems

5. Closure Theorem 1. (L. Cesari) Let A be a closed subset of $E_1 \times E_n$, let $U(t, x)$ be a closed subset of E_m for every $(t, x) \in A$, let $f(t, x, u) = (f_1, \ldots, f_n)$ be a continuous vector function on M into E_n, and let $Q(t, x) = f(t, x, U(t, x))$ be a closed convex subset of E_n for every $(t, x) \in A$. Assume that $U(t, x)$ has property (U) in A, and that $Q(t, x)$ has property (Q) in A. Let $x_k(t)$, $t_{1k} \leq t \leq t_{2k}$, $k = 1, 2, \ldots$, be a sequence of trajectories, which is convergent in the metrics ρ toward an absolutely continuous function $x(t)$, $t_1 \leq t \leq t_2$. Then $x(t)$ is a trajectory.

Remark. If we assume that $U(t, x)$ is compact for every $(t, x) \in A$, and that $U(t, x)$ is an uppersemicontinuous function of (t, x) in A, then the set $Q(t, x)$ has the same property, $U(t, x)$ has property (U), $Q(t, x)$ has property (Q), and closure theorem I reduces to one of A. F. Filippov [2] (not explicitely stated in [2] but contained in the proof of his existence theorem for the Pontryagin problem with $U(t, x)$ always compact).

Proof of Closure Theorem I. The vector functions

$$\varphi(t) = x'(t), \quad t_1 \leq t \leq t_2,$$

(8)

$$\varphi_k(t) = x'_k(t) = f(t, x_k(t), u_k(t)), \quad t_{1k} \leq t \leq t_{2k}, \quad k = 1, 2, \ldots,$$

are defined almost everywhere and are L-integrable. We have to prove that $(t, x(t)) \in A$ for every $t_1 \leq t \leq t_2$, and that there is a measurable control function $u(t)$, $t_1 \leq t \leq t_2$, such that

(9) $$\varphi(t) = x'(t) = f(t, x(t), u(t)), \quad u(t) \in U(t, x(t)),$$

for almost all $t \in [t_1, t_2]$.

First, $\rho(x_k, x) \to 0$ as $k \to \infty$, hence , $t_{1k} \to t_1$, $t_{2k} \to t_2$. If $t \in (t_1, t_2)$, or $t_1 < t < t_2$, then $t_{1k} < t < t_{2k}$ for all k sufficiently large and $(t, x_k(t)) \in A$. Since $x_k(t) \to x(t)$ as $k \to \infty$ and A is closed, we conclude that $(t, x(t)) \in A$ for every $t_1 < t < t_2$. Since $x(t)$ is continuous, and hence continuous at t_1 and t_2, we conclude that $(t, x(t)) \in A$ for every $t_1 \leq t \leq t_2$.

For almost all $t \in [t_1, t_2]$ the derivative $x'(t)$ exists and is finite. Let t_0 be such a point with $t_1 < t_0 < t_2$. Then there is a $\sigma > 0$ with $t_1 < t_0 - \sigma < t_0 + \sigma < t_2$, and, for some k_0 and all $k \geq k_0$, also $t_{1k} < t_0 - \sigma < t_0 + \sigma < t_{2k}$. Let $x_0 = x(t_0)$.

We have $x_k(t) \to x(t)$ uniformly in $[t_0 - \sigma, t_0 + \sigma]$ and all functions $x(t)$, $x_k(t)$ are continuous in the same interval. Thus, they are equicontinuous in $[t_0 - \sigma, t_0 + \sigma]$. Given $\epsilon > 0$, there is a $\delta > 0$ such that $t, t' \in [t_0 - \sigma, t_0 + \sigma]$, $|t - t'| \leq \delta$, $k \geq k_0$, implies

$$|x(t) - x(t')| \leq \epsilon/2, \quad |x_k(t) - x_k(t')| \leq \epsilon/2.$$

We can assume $0 < \delta < \sigma$, $\delta \leq \epsilon$. For any h, $0 < h \leq \delta$, let us consider the averages

(10)
$$m_h = h^{-1} \int_0^h \varphi(t_0 + s)ds = h^{-1} [x(t_0 + h) - x(t_0)],$$

$$m_{hk} = h^{-1} \int_0^h \varphi_k(t_0 + s)ds = h^{-1} [x_k(t_0 + h) - x_k(t_0)].$$

Given $\eta > 0$ arbitrary, we can fix h, $0 < h \leq \delta < \sigma$, so small that

(11) $$|m_h - \varphi(t_o)| \leq \eta .$$

Having so fixed h, let us take $k_1 \geq k_o$ so large that

(12) $$|m_{hk} - m_h| \leq \eta , \quad |x_k(t_o) - x(t_o)| \leq \epsilon/2$$

for all $k \geq k_1$. This is possible since $x_k(t) \to x(t)$ as $k \to \infty$ both at $t = t_o$ and $t = t_o + h$. Finally, for $0 \leq s \leq h$, $k \geq k_1$,

$$|x_k(t_o + s) - x(t_o)| \leq |x_k(t_o + s) - x_k(t_o)| + |x_k(t_o) - x(t_o)| \leq \epsilon/2 + \epsilon/2 = \epsilon,$$

$$|(t_o + s) - t_o| \leq h \leq \delta \leq \epsilon ,$$

$$f(t_o + s, x_k(t_o + s), u_k(t_o + s)) \in Q(t_o + s, x_k(t_o + s)).$$

Hence , by the definition of $Q(t_o, x_o, 2\epsilon)$, also

$$\varphi_k(t_o + s) = f(t_o + s, x_k(t_o + s), u_k(t_o + s)) \in Q(t_o, x_o, 2\epsilon).$$

The second integral relation (10) shows that we have also

$$m_{1k} \in \text{cl co } Q(t_o, x_o, 2\epsilon) ,$$

since the latter is a closed convex set. Finally by relations (11) and (12) , we deduce

$$|\varphi(t_o) - m_{hk}| \leq |\varphi(t_o) - m_h| + |m_h - m_{hk}| \leq 2\eta,$$

and hence

$$\varphi(t_o) \in [\mathrm{cl}\ \mathrm{co}\ Q(t_o, x_o, 2\epsilon)]_{2\eta} .$$

Here $\eta > 0$ is an arbitrary number, and the set in brackets is closed. Hence,

$$\varphi(t_o) \in \mathrm{cl}\ \mathrm{co}\, Q(t_o, x_o, 2\epsilon) ,$$

and this relation holds for every $\epsilon > 0$. By property (Q) we have

$$\varphi(t_o) \in \bigcap_{\epsilon} \mathrm{cl}\ \mathrm{co}\ Q(t_o, x_o, 2\epsilon) = Q(t_o, x_o) ,$$

where $x_o = x(t_o)$, and $Q(t_o, x_o) = f(t_o, x_o, U(t_o, x_o))$. This relation implies that there are points $\bar{u} = \bar{u}(t_o) \in U(t_o, x_o)$ such that

(13) $$\varphi(t_o) = f(t_o, x(t_o), \bar{u}(t_o)) .$$

This holds for almost all $t_o \in [t_1, t_2]$, that is, for all t of a measurable set $I \subset [t_1, t_2]$ with meas $I = t_2 - t_1$. If we take $I_o = [t_1, t_2] - I$, then meas $I_o = 0$. Hence, there is at least one function $\bar{u}(t)$, defined almost everywhere in $[t_1, t_2]$, for which relation (13) holds a.e. in $[t_1, t_2]$. We have to prove that there is at least cne such function which is measurable. For every $t \in I$, let $P(t)$ denote the set

$$P(t) = [u \mid u \in U(t, x(t)),\ \varphi(t) = f(t, x(t), u)] \subset U(t, x(t)) \subset E_m .$$

We have proved that $P(t)$ is not empty.

For every integer $\lambda = 1, 2, \dots$, there is a closed subset C_λ of $I, C_\lambda \subset I \subset [t_1, t_2]$, with meas $C_\lambda > \max [0, t_2 - t_1 - 1/\lambda]$, such that $\varphi(t)$ is continuous on C_λ. Let W_λ be the set

$$W_\lambda = [(t,u) | t \in C_\lambda \ , \ u \in P(t)] \subset E_1 \times E_m \ .$$

Let us prove that the set W_λ is closed. Indeed, if $(\bar{t}, \bar{u})$ is a point of accumulation of A_λ, then there is a sequence (t_s, u_s), $s = 1, 2, \ldots$, with $(t_s, u_s) \in W_\lambda$, $t_s \to \bar{t}$, $u_s \to \bar{u}$. Then, $t_s \in C_\lambda$ and $\bar{t} \in C_\lambda$ since C_λ is closed. Also $x(t_s) \to x(\bar{t})$, $\varphi(t_s) \to \varphi(\bar{t})$, and since $(t_s, x(t_s)) \in A$, $\varphi(t_s) = f(t_s, x(t_s), u(t_s))$, $(t_s, x(t_s), u(t_s)) \in M$, we have also $(\bar{t}, x(\bar{t})) \in A$, $(t, x(\bar{t}), \bar{u}) \in M$, because A and M are closed, and $\varphi(\bar{t}) = f(\bar{t}, x(\bar{t}), \bar{u})$ because f is continuous. Thus, $\bar{u} \in P(\bar{t})$, and $(\bar{t}, \bar{x}) \in W_\lambda$. This proves that W_λ is a closed set.

For every integer ℓ let $W_{\lambda\ell}$, $P_\ell(t)$ be the sets

$$W_{\lambda\ell} = [(t,u) \mid (t,u) \in W_\lambda, \ |u| \leq \ell] \subset W_\lambda \subset E_1 \times E_m ,$$

$$P_\ell(t) = [u \mid u \in P(t), \ |u| \leq \ell] \subset P(t) \subset U(t, x(t)) \subset E_m ,$$

$$C_{\lambda\ell} = [t \mid (t,u) \in W_{\lambda\ell} \text{ for some } u] \subset C_\lambda \subset I \subset [t_1, t_2] .$$

Obviously, $W_{\lambda\ell}$ is compact, and so is $C_{\lambda\ell}$ as its projection on the t-axis. Also $\bigcup_\ell C_{\lambda\ell} = C_\lambda$, and $W_{\lambda\ell}$ is the set of all (t,u) with $t \in C_{\lambda\ell}$, $u \in P_\ell(t)$. Thus, for $t \in C_{\lambda\ell}$, $P_\ell(t)$ is a compact subset of $U(t, x(t))$.

For $t \in C_{\lambda\ell}$, and ℓ large enough the set $P_\ell(t)$ is a nonempty compact subset of all $u = (u^1, \ldots, u^m) \in U(t, x(t))$ with $f(t, x(t), u) = \varphi(t)$ and $|u| \leq \ell$. Let P_1 be the subset of P_ℓ with u^1 minimum; let P_2 be the subset of P_1 with u^2 minimum, ..., let P_m be the subset of P_{m-1} with u^m minimum. Then P_m is a single point $u = u(t) \in U(t, x(t))$ with $u(t) = (u^1, \ldots, u^m)$, $t \in C_{\lambda\ell}$, $|u(t)| \leq \ell$, and $f(t, x(t), u(t)) = \varphi(t)$. Let us prove that $u(t)$, $t \in C_{\lambda\ell}$, is measurable. We shall prove this by induction on the coordinates. Let us assume that $u^1(t), \dots, u^{s-1}(t)$ have been proved to be measurable on $C_{\lambda\ell}$ and let us prove that $u^s(t)$ is measurable. For s=1 nothing is assumed, and the argument below proves that $u^1(t)$ is measurable. For every integer j there are closed subsets $C_{\lambda\ell j}$ with $C_{\lambda\ell j} \subset C_{\lambda\ell}$, $C_{\lambda\ell j} \subset C_{\lambda\ell, j+1}$, meas $C_{\lambda\ell j} > \max [0, \text{meas } C_{\lambda\ell} - 1/j]$, such that $u^1(t), \dots, u^{s-1}(t)$ are continuous on $C_{\lambda\ell j}$. The function $\varphi(t)$ is already continuous on C_λ and hence $\varphi(t)$ is continuous on every set $C_{\lambda\ell}$ and $C_{\lambda\ell j}$. Let us prove that $u^s(t)$ is measurable on $C_{\lambda\ell j}$. We have only to prove that, for every real a, the set of all $t \in C_{\lambda\ell j}$ with $\varphi^s(t) \leq a$ is closed. Suppose that this is not the case. Then there is a sequence of points $t_k \in C_{\lambda\ell j}$ with $u^s(t_k) \leq a$, $t_k \to \bar{t} \in C_{\lambda\ell j}$, $u^s(\bar{t}) > a$. Then $\varphi(t_k) \to \varphi(\bar{t})$, $u^\alpha(t_k) \to u^\alpha(\bar{t})$ as $k \to \infty$, $\alpha = 1, \dots, s-1$. Since $|u^\beta(t_k)| \leq \ell$ for all k and $\beta = s, s+1, \dots, m$, we can select a subsequence, say still $[t_k]$, such that $u^\beta(t_k) \to \tilde{u}^\beta$ as $k \to \infty$, $\beta = s, s+1, \dots, m$, for some real numbers $\tilde{u}^\beta$. Then $t_k \to \bar{t}$, $x(t_k) \to x(\bar{t})$, $u(t_k) \to \tilde{u}$, where

$$\tilde{u} = (u^1(\bar{t}), \dots, u^{s-1}(\bar{t}), \tilde{u}^s, \dots, \tilde{u}^m).$$

Then, given any number $\eta > 0$, we have

$$u(t_k) \in U(t_k, x(t_k)) \subset \text{cl } U(\bar{t}, c(\bar{t}), \eta)$$

for all k sufficiently large, and as $k \to \infty$, also

$$\tilde{u} \in \text{cl } U(\bar{t}, x(\bar{t}), \eta).$$

By property (U) we have

$$\tilde{u} \in \bigcap_\eta \text{cl } U(\bar{t}, x(\bar{t}), \eta) = U(\bar{t}, x(\bar{t})).$$

On the other hand $\varphi(t_k) = f(t_k, x(t_k), u(t_k))$, $u^s(t_k) \leq a$, yield as $k \to \infty$,

(14) $$\varphi(\bar{t}) = f(\bar{t}, x(\bar{t}), \tilde{u}), \quad \tilde{u}^s < a,$$

while $\bar{t} \in C_{\lambda \ell}$ implies

(15) $$\varphi(\bar{t}) = f(\bar{t}, x(\bar{t}), u(\bar{t})), \quad u^s(\bar{t}) > a .$$

Relations (14) and (15) are contradictory, because of the minimum property with which $u^s(\bar{t})$ has been chosen. Thus, $u^s(t)$ is measurable on $C_{\lambda \ell j}$ for every j, and then $u^s(t)$ is also measurable on $C_{\lambda \ell}$. By induction all components $u^1(t), \dots, u^m(t)$ of u(t) are measurable on $C_{\lambda \ell}$, hence, u(t) is measurable on $C_{\lambda \ell}$. Since $\bigcup_\ell C_{\lambda \ell} = C_\lambda$, meas $C_\lambda >$ meas $I - 1/\lambda$, we conclude that u(t) is measurable on every set C_λ and hence on I, with meas $I = t_2 - t_1$. Thus u(t) is defined a.e., on $[t_1, t_2]$, $u(t) \in U(t, x(t))$ and $f(t, x(t), u(t)) = \varphi(t)$ a.e. on $[t_1, t_2]$. Closure Theorem I is thereby proved.

<u>Remark</u>. The last part of the proof of Closure Theorem I concerning the existence of al least one measurable u(t) is a modification for U(t, x) closed and (satisfying condition (U)) of the analogous argument of A.F. Filippov [2] for the case where U(t, x) is an upper semicontinuous compact subset of a Euclidean space E_m. A different argument-again concerning only the last part of the proof-has been devised by C. Castaing [C.R. Acad. Sci. Paris, 262, 1966, 409-411] for the case where U depends on t only and is an upper semicontinuous compact subset of any separable metric space E.

6. Another closure theorem.

Let us denote by $y = (x^1, \dots, x^s)$ the s-vector made up of certain components, say $x^1, \dots, x^s$, $0 \leq n$, of $x = (x^1, \dots, x^n)$, and by z the complementary (n-s)-vector $z = (x^{s+1}, \dots, x^n)$ of x, so that $x = (y, z)$. Let us assume that $f(t, y, u)$ depends only on the coordinates $x^1, \dots, x^s$ of x. If $x(t)$, $t_1 \leq t \leq t_2$, is any vector function, we shall denote by $x(t) = [y(t), z(t)]$ the corresponding decomposition of $x(t)$ in its coordinates $y(t) = (x^1, \dots, x^s)$ and $z(t) = (x^{s+1}, \dots, x^n)$.

We shall denote by A_o a closed subset of points $(t, x^i, \dots, x^s)$, that is, a closed subset of the ty-space $E_1 \times E_s$, and let $A = A_o \times E_{n-s}$. Thus, A is a closed subset of the tx-space $E_1 \times E_n$.

Closure Theorem II. (L. Cesari [1e]). Let A_o be a closed subset of the ty-space $E_1 \times E_s$, and then $A = A_o \times E_{n-s}$ is a closed subset of the tx-space $E_1 \times E_n$. Let $U(t, y)$ denote a closed subset of E_m for every $(t, y) \in A_o$, let M_o be the set of all $(t, y, u) \in E_{1+s+m}$ with $(t, y) \in A_o$, $u \in U(t, y)$, and let $f(t, y, u) = (f_1, \dots, f_n)$ be a continuous vector function from M into E_n. Let $Q(t, y) = f(t, y, U(t, y))$ be a closed convex subset of E_n for every $(t, y) \in A_o$. Assume that $U(t, y)$ has property (U) in A_o and that $Q(t, y)$ has property (Q) in A_o. Let $x_k(t)$, $t_{1k} \leq t \leq t_{2k}$, $k = 1, 2, \dots$, be a sequence of trajectories, $x_k(t) = (y_k(t), z_k(t))$, for which we assume that the s-vector $y_k(t)$ converges in the ρ-metric toward an AC vector function $y(t)$, $t_1 \leq t \leq t_2$, and that the (n-k)-vector $z_k(t)$ converges (pointwise) for almost all $t_1 < t < t_2$, toward a vector $z(t)$ which admits of a decomposition $z(t) = Z(t) + S(t)$ where $Z(t)$ is an AC vector function in $[t_1, t_2]$, and $S'(t) = 0$ a.e. in $[t_1, t_2]$ (that is, $S(t)$ is a singular function). Then, the AC vector $X(t) = [y(t), Z(t)]$, $t_1 \leq t \leq t_2$, is a trajectory.

Remark. For $s = n$, this theorem reduces to closure theorem I

Proof of Closure Theorem II. The vector functions

$$\varphi(t) = X'(t) = (y'(t), Z'(t)), \quad t_1 \leq t \leq t_2,$$

$$(16) \quad \varphi_k(t) = x'_k(t) = (y'_k(t), z'_k(t), Z'_k(t)) = f(t, y_k(t), u_k(t)), \quad t_{1k} \leq t \leq t_{2k},$$

$$k = 1, 2, \dots,$$

are defined almost everywhere and are L-integrable. We have to prove that $[t, y(t), Z(t)] \in A$ for every $t_1 \leq t \leq t_2$, and that there is a measurable control function $u(t)$, $t_1 \leq t \leq t_2$ such that

$$(17) \quad \varphi(t) = X'(t) = (y'(t), Z'(t)) = f(t, y(t), u(t)),$$

$$u(t) \in U(t, y(t)),$$

for almost all $t \in [t_1, t_2]$.

First, $\rho(y_k, y) \to 0$ as $k \to 0$; hence $t_{1k} \to t_1$, $t_{2k} \to t_2$. If $t \in (t_1, t_2)$, or $t_1 < t < t_2$, then $t_{1k} < t < t_{2k}$ for all k sufficiently large, and $(t, y_k(t) \in A_o$. Since $y_k(t) \to y(t)$ as $k \to \infty$ and A_o is closed, we conclude that $(t, y(t)) \in A_o$ for every $t_1 < t < t_2$, and finally $(t, y(t), Z(t)) \in A_o \times E_{n-s}$, or $(t, X(t)) \in A$, $t_1 \leq t \leq t_2$.

For almost all $t \in [t_1, t_2]$ the derivative $X'(t) = [y'(t), Z'(t)]$ exists and is finite, $S'(t)$ exists and $S'(t) = 0$, and $z_k(t) \to z(t)$. Let t_o be such a point with $t_1 < t_o < t_2$. Then there is a $\sigma > 0$ with

$t_1 < t_o - \sigma < t_o + \sigma < t_2$, and, for some k_o and all $k \geq k_o$, also $t_{1k} < t_o - \sigma < t_o + \sigma < t_{2k}$. Let $x_o = X(t_o) = (y_o, Z_o)$, or $y_o = y(t_o)$, $Z_o = Z(t_o)$. Let $z_o = z(t_o)$, $S_o = S(t_o)$. We have $S'(t_o) = 0$, hence $z'(t_o)$ exists and $z'(t_o) = Z'(t_o)$. Also, we have $z_k(t_o) \rightarrow z(t_o)$.

We have $y_k(t) \rightarrow y(t)$ uniformly in $[t_o - \sigma, t_o + \sigma]$, and all functions $y(t)$, $y_k(t)$ are continuous in the same interval. Thus they are equicontinuous in $[t_o - \sigma, t_o + \sigma]$. Given $\epsilon > 0$, there is a $\delta > 0$ such that $t, t' \in [t_o - \sigma, t_o + \sigma]$, $|t - t'| \leq \delta$, $k \geq k_o$, implies

$$|y(t) - y(t')| \leq \epsilon/2 , \quad |y_k(t) - y_k(t')| \leq \epsilon/2 .$$

We can assume $0 < \delta < \sigma$, $\delta \leq \epsilon$. For any h, $0 < h \leq \delta$, let us consider the averages

$$m_h = h^{-1} \int_0^h \varphi(t_o + s)ds = h^{-1} [X(t_o + h) - X(t_o)] ,$$

(18)

$$m_{hk} = h^{-1} \int_0^h \varphi_k(t_o + s)ds = h^{-1}[x_k(t_o + h) - x_k(t_o)]$$

where $X = (y, Z)$, $x_k = (y_k, z_k)$.

Given $\eta > 0$ arbitrary, we can fix $h, 0 < h \leq \delta < \sigma$, so small that

$$|m_h - \varphi(t_o)| \leq \eta ,$$

$$|S(t_o + h) - S(t_o)| < \eta h/4 .$$

This is possible since $h^{-1} \int_0^h \varphi(t_o + s)\, ds \rightarrow \varphi(t_o)$ and $[S(t_o + h) - S(t_o)] h^{-1} \rightarrow 0$ as $h \rightarrow 0+$. Also, we can choose h, in such a way that $z_k(t_o + h) \rightarrow z(t_o)$ as $k \rightarrow +\infty$. This is possible since $z_k(t) \rightarrow z(t)$ for almost all $t_1 < t < t_2$.

Having so fixed h, let us take $k_1 \geq k_o$ so large that

$$|y_k(t_o) - y(t_o)|, |y_k(t_o + h) - y(t_o + h)| \leq \min[\eta h/4, \epsilon/2],$$

$$|z_k(t_o) - z(t_o)|, |z_k(t_o + h) - z(t_o + h)| \leq \eta h/8 .$$

This is possible since $y_k(t) \to y(t)$, $z_k(t) \to z(t)$ both at $t = t_o$ and $t = t_o + h$. Then we have

$$|h^{-1}[y_k(t_o + h) - y_k(t_o)] - h^{-1}[y(t_o + h) - y(t_o)]|$$

$$\leq |h^{-1}[y_k(t_o + h) - y(t_o + h)]| + |h^{-1}[y_k(t_o) - y(t_o)]|$$

$$\leq h^{-1}(\eta h/4) + h^{-1}(\eta h/4) = \eta/2 .$$

Analogously, since $z = Z + S$, we have

$$|h^{-1}[z_k(t_o + h) - z_k(t_o)] - h^{1}[Z(t_o + h) - Z(t_o)]|$$

$$= |h^{-1}[z_k(t_o + h) - z_k(t_o)] - h^{-1}[z(t_o + h) - z(t_o)] + h^{-1}[S(t_o + h) - S(t_o)]|$$

$$\leq |h^{-1}[z_k(t_o + h) - z(t_o + h)]| + |h^{-1}[z_k(t_o) - z(t_o)]| + |h^{-1}[S(t_o + h) - S(t_o)]|$$

$$\leq h^{-1}(\eta h/8) + h^{-1}(\eta h/8) + h^{-1}(\eta h/4) = \eta/2.$$

Finally, we have

$$|m_{hk} - m_h| = |h^{-1}[x_k(t_o + h) - x_k(t_o)] - h^{-1}[X(t_o + h) - X(t_o)]|$$

$$\leq |h^{-1}[y_k(t_o + h) - y_k(t_o)] - h^{-1}[y(t_o + h) - y(t_o)]| +$$

$$+ |h^{-1}[z_k(t_o + h) - z_k(t_o)] - h^{-1}[Z(t_o + h) - Z(t_o)]|$$

$$\leq \eta/2 + \eta/2 = \eta .$$

We conclude that for the chosen value of h, $0 < h \leq \delta < \sigma$, and every $k \geq k_1$ we have

(19) $$|m_h - \varphi(t_o)| \leq \eta , \ |m_{hk} - m_h| \leq \eta , \ |y_k(t_o) - y(t_o)| < \epsilon/2 .$$

For $0 \leq s \leq h$ we have now

$$|y_k(t_o + s) - y(t_o)| \leq |y_k(t_o + s) - y_k(t_o)| + |y_k(t_o) - y(t_o)| \leq \epsilon/2 + \epsilon/2 = \epsilon ,$$

$$|(t_o + s) - t_o| \leq h \leq \delta \leq \epsilon ,$$

$$f(t_o + s, y_k(t_o + s), u_k(t_o + s)) \in Q(t_o + s, y_k(t_o + s)) .$$

Hence, by definition of $Q(t_o, y_o, 2\epsilon)$, also

$$\varphi_k(t_o + s) = f(t_o + s, y_k(t_o + s), u_k(t_o + s)) \in Q(t_o, y_o, 2\epsilon) .$$

The second integral relation (18) shows that we have also

$$m_{hk} \in \mathrm{cl\ co}\ Q(t_o, y_o, 2\epsilon) ,$$

since the latter is a closed convex set. Finally, by relations (19), we deduce

$$|\varphi(t_o) - m_{hk}| \leq |\varphi(t_o) - m_h| + |m_h - m_{hk}| \leq 2\eta ,$$

and hence

$$\varphi(t_o) \in [\mathrm{cl\ co}\ Q(t_o, y_o, 2\epsilon)]_{2\eta}$$

Here $\eta > 0$ is an arbitrary number, and the set in brackets is closed. Hence

$$\varphi(t_o) \in \mathrm{cl\ co}\, Q(t_o, y_o, 2\epsilon) ,$$

and this relation holds for every $\epsilon > 0$. By property (Q) we have

$$\varphi(t_o) \in \bigcap_\epsilon \mathrm{cl\ co}\, Q(t_o, y_o, 2\epsilon) = Q(t_o, y_o) ,$$

where $y_o = y(t_o)$ and $Q(t_o, y_o) = f(t_o, y_o, U(t_o, y_o))$. This relation implies that there are points $\bar{u} = \bar{u}(t_o) \in U(t_o, y_o)$ such that

$$\varphi(t_o) = f(t_o, y(t_o), \bar{u}(t_o)) .$$

This holds for almost all $t_o \in [t_1, t_2]$. Hence , there is at least one function $\bar{u}(t)$, defined a.e. in $[t_1, t_2]$ for which relation (17) holds a.e. in $[t_1, t_2]$. We have to prove that there is at least one such function which is measurable. The proof is exactly as the one for closure theorem I, where we write y, y_k instead of x, x_k, and will not be repeated here. Closure theorem II is thereby proved.

Lecture 4. Existence theorems for usual solutions of Lagrange problems.

7. Notations . Again we shall use the notations of no. 1. It will be convenient to write the problem in a slightly different form. First we introduce the auxiliary variable x^o satisfying the differential equation and initial value

$$dx^o/dt = f_o(t, x(t), u(t)), \quad x^o(t_1) = 0, \quad x^o(t) \text{ AC in } [t_1, t_2].$$

Then

$$x^o(t_2) = \int_{t_1}^{t_2} f_o(t, x(t), u(t))dt = I[x, u].$$

If now we denote by $\tilde{x}$ the (n+1) - vector $\tilde{x} = (x^o, x^1, \dots, x^n)$, and by $\tilde{f}(t, x, u)$ the (n+1) - vector function $\tilde{f}(t, x, u) = (f_o, f_1, \dots, f_n)$, then the problem of minimum discussed in no. 1 reduces to the determination of a pair $[\tilde{x}(t), u(t)]$, $t_1 \le t \le t_2$, satisfying the differential system

$$d\tilde{x}/dt = \tilde{f}(t, x(t), u(t)) \quad \text{a.e. in } [t_1, t_2],$$

the boundary conditions

$$(t_1, x(t_1), t_2, x(t_2)) \in B, \quad x^o(t_1) = 0,$$

and the constraints

$$(t, x(t)) \in A, \qquad u(t) \in U(t, x(t)),$$

for which $x^o(t_2)$ has its minimum value. Here $x(t) = (x^1, \dots, x^n)$, $\tilde{x}(t) = (x^o, x^1, \dots, x^n)$, and the present formulation corresponds to a transformation of the Lagrange type problem of no. 1 into a problem of the Mayer type.

We shall now consider for every $(t,x) \in A$ the sets $Q(t,x)$, $\tilde{Q}(t,x)$, $\tilde{\tilde{Q}}(t,x)$ defined as follows:

$$Q(t,x) = f(t,x,U(t,x)) = [z \mid z = f(t,x,u),\ u \in U(t,x)] \subset E_n,$$

$$(20) \qquad \tilde{Q}(t,x) = \tilde{f}(t,x,U(t,x)) = [\tilde{z} = (z^o, z) \mid \tilde{z} = \tilde{f}(t,x,u), u \in U(t,x)]$$

$$= [\tilde{z} = (z^o, z) \mid z^o = f_o(t,x,u),\ z = f(t,x,u),\ u \in U(t,x)] \subset E_{n+1},$$

$$\tilde{\tilde{Q}}(t,x) = [\tilde{z} = (z^o, z) \mid z^o \geq f_o(t,x,u), z = f(t,x,u), u \in U(t,x)] \subset E_{n+1}.$$

The main hypothesis of the existence theorems which we shall state and prove below is that the set $\tilde{\tilde{Q}}(t,x)$ be <u>convex</u> for every $(t,x) \in A$.

For free problems that is, for $m=n$, $U = E_n$, $f=u$, the sets Q $\tilde{Q}, \tilde{\tilde{Q}}$ thought of as subsets of the z^o u-space are

$$Q(t,x) = E_n,$$

$$\tilde{Q}(t,x) = [z = (z^o, u) \mid z^o = f^o(t,x,u),\ u \in E_n] \subset E_{n+1},$$

$$\tilde{\tilde{Q}}(t,x) = [z = (z^o, u) \mid z^o \geq f_o(t,x,u),\ u \in E_n] \subset E_{n+1}$$

Thus, the convexity of $\tilde{\tilde{Q}}$ reduces to the usual convexity cnndition of $f_o(t,x,u)$ as a function of u in E_n – a condition which is familiar in the calculus of variations for free problems.

We mention here that a function $\phi(u)$, $u \in E_n$, is said to be convex in u, provided u, $v \in E_n$, $0 \leq \alpha \leq 1$, implies $\phi(\alpha u + (1-\alpha)v) \leq$ $\leq \alpha\phi(u) + (1-\alpha)\phi(v)$.

<u>8. Statement of the first existence theorem</u>.

Existence Theorem 1 (for Lagrange problems with of without unilateral constraints) (L. Cesari [1e]. Let A be any compact subset of the tx-space $E_1 \times E_n$, and for every $(t,x) \in A$ let $U(t,x)$ be a closed subset of the u-space E_m. Let M be the set of all (t,x,u) with $(t,x) \in A$, $u \in U(t,x)$, and let $\tilde{f}(t,x,u) = (f_o, f) = (f_o, f_1, \dots, f_n)$ be a continuous vector function on M. For every $(t,x) \in A$ let

$$\tilde{\tilde{Q}}(t,x) = \left\{ \tilde{z} = (z^o, z) \mid z^o \geq f_o(t,x,u),\ z = f(t,x,u),\ u \in U(t,x) \right\} \subset E_{n+1}$$

be a convex subset of E_{n+1}. Let us assume that $U(t,x)$ satisfies condition (U) in A, and that $\tilde{\tilde{Q}}(t,x)$ satisfies condition (Q) in A. Let $\varphi(\zeta)$, $0 \leq \zeta < +\infty$, be a given continuous function of ζ satisfying the relation $\varphi(\zeta)/\zeta \to +\infty$ as $\zeta \to +\infty$, and assume that $f_o(t,x,u) \geq \varphi(|u|)$ for all $(t,x,u) \in M$. Also, let C;D be constants and assume that $|f(t,x,u)| \leq C + D|u|$ for all $(t,x,u) \in M$. Let B be a closed subset of the $t_1 x_1 t_2 x_2$-space E_{2n+2}. Then the cost functional $I[x,u]$ has an absolute minimum in any nonempty complete the class Ω of admissible pairs $x(t)$, $u(t)$.

If A is not compact, but closed and contained in a slab $\left\{ t_o \leq t \leq T,\ -\infty < x^i < +\infty,\ i = 1, \dots, n,\ t_o, T \text{ finite} \right\}$, then Existence Theorem 1 still holds under the additional hypotheses that

(a) $x^1 f_1 + \dots + x^n f_n \leq N[(x^1)^2 + \dots + (x^n)^2 + 1]$ for all (t,x,u) M and some constant $N > 0$, and

b) each trajectory $x(t)$ of the class Ω contains at least one point $(t^*, x(t^*))$ on a compact subset P of A (for instance, the initial point $(t_1, x(t_1))$ is fixed, or the end point is fixed).

If A is not compact, nor contained in a slab as above, but A is closed, then Existence Theorem 1 still holds under the additional hypotheses (a), (b), and (c):

(c) $f_o(t,x,u) \geqq \mu > 0$ for all $(t,x,u) \in M$ with $|t| \geqq R$ and for some constants $\mu > 0$, $R > o$.

Also, condition (a) be replaced by the following condition (d) :

(d) $f_o(t,x,u) \geqq E\,|f(t,x,u)|$ for all $(t,x,u) \in M$ with $|x| \geqq F$ and for some constants $E > 0$ and $F \geqq 0$.

Condition (b) is certainly verified if for instance the projection B_1 of B on the $t_1 x_1$ space is compact, in particular, if the initial points $(t_1, x(t_1))$ of the trajectories x(t) of Ω belong to a compact subset of the same space. The same holds if the projection B_2 on the $t_2\, x_2$-space is compact, in particular if the endpoints $(t_2, x(t_2))$ of the trajectories x(t) of Ω belong to a compact subset of the same space.

Finally, for A not compact but closed, the conditions $f_o \geqq \varphi(|u|)$, $|f| \leqq C + D\,|u|$ above can be replaced by the following condition (g) :

(g) for every compact subset A_o of A there are functions φ_o as above and constants $C_o \geqq 0$, $D_o \geqq 0$ (all may depend on A_o) such that $f_o \geqq \varphi_o(|u|)$, $|f| \leqq C_o + D_o\,|u|$ for all $(t,x,u) \in M$ with $(t,x), \in A_o$

<u>Proof of Existence Theorem I</u>. We have $\varphi(\zeta) \geq -M_o$ for some number $M_o \geq 0$, hence $\varphi(\zeta) + M_o \geq 0$ for all $\zeta \geq 0$, and $f_o(t,x,u) + M_o \geq 0$ for all $(t,x,u) \in M$. Let D be the diameter of A Then for every pair x(t), u(t), $t_1 \leq t \leq t_2$, of Ω we have .

$$I[x, u] = \int_{t_1}^{t_2} f_o\,dt \geq \int_{t_1}^{t_2} \varphi(|u|)\,dt \geq -DM_o > -\infty.$$

Let $i = \text{Ifn}\ I[x,u]$, where Inf is taken for all pairs $x(t)$, $u(t)$ in $\in \Omega$. Then i is finite.

Let $x_k(t)$, $u_k(t)$, $t_{1k} \le t \le t_{2k}$, $k = 1, 2, \dots$ be a sequence of admissible pairs all in Ω, such that $I[x_k, u_k] \to i$ as $k \to \infty$. We may assume

$$i \le I[x_k, u_k] = \int_{t_{1k}}^{t_{2k}} f_o(t, x_k(t), u_k(t))dt \le i + k^{-1} \le i+1, k=1,2,\dots$$

Let us prove that the AC vector functions $x_k(t)$, $t_{1k} \le t \le t_{2k}$, $k = 1, 2, \dots$, are equiabsolutely continuous. Let $\epsilon > 0$ be any given number, and let $\sigma = 2^{-1}\epsilon\ (DM_o + |i| + 1)^{-1}$.

Let $N > 0$ be a number such that $\varphi(\zeta)/\zeta > 1/\sigma$ for $\zeta \ge N$. Let E be any measurable subset of $[t_{1k}, t_{2k}]$ with meas $E < \eta = \epsilon/2N$. Let E_1 be the subset of all $t \in E$ where $u_k(t)$ is finite and $|u_k(t)| \le N$, and let $E_2 = E - E_1$. Then $|u_k(t)| \le N$ in E_1, and $\varphi(|u_k|)/|u_k| \ge 1/\sigma$, or $u_k \le \sigma\varphi(|u_k|)$, a.e. in E_2. Hence

$$\begin{aligned}\int_E |u_k(t)|dt &= \left(\int_{E_1} + \int_{E_2}\right)|u_k(t)|\,dt \\ &\le N \text{ meas } E_1 + \sigma\int_{E_2}\varphi(|u_k(t)|)\,dt \\ &\le N \text{ meas } E + \sigma\int_{E_2}[\varphi(|u_k(t)|) + M_o]\,dt \\ &\le N\eta + \sigma\int_{t_{1k}}^{t_{2k}}[\varphi(|u_k(t)|) + M_o]\,dt \\ &\le N\eta + \sigma\int_{t_{1k}}^{t_{2k}}[f_o(t, x_k(t), u_k(t)) + M_o]\,dt\end{aligned}$$

$$\leq N\eta + \sigma(DM_o + |i| + 1)$$

(22) $$= \epsilon/2 + \epsilon/2 = \epsilon .$$

This proves that the vector functions $u_k(t)$, $t_{1k} \leq t \leq t_{2k}$, $k = 1, 2, \ldots$, are equabsolutely integrable . From here we deduce

$$\int_E |x'_k(t)|\, dt = \int_E |f(t, x_k(t), u_k(t)|\, dt \leq \int_E [A+B|u_k(t)|]\, dt$$

$$\leq C \text{ meas } E + D \int_E |u_k(t)|\, dt ,$$

and this proves the equiabsolute continuity of the vector functions $x_k(t)$, $t_{1k} \leq t \leq t_{2k}$, $k = 1, 2, \ldots$.

Now let us consider the sequence of AC scalar functions $x^o_k(t)$ defined by

(23) $$x^o_k(t) = \int_{t_{1k}}^{t} f_o(\tau, x_k(\tau), u_k(\tau))\, d\tau , \quad t_{1k} \leq t \leq t_{2k} .$$

Then ,

$$x^o_k(t_{1k}) = 0, \quad x^o_k(t_{2k}) = I[x_k, u_k] \to i \text{ as } k \to +\infty ,$$

and

$$i \leq x^o_k(t_{2k}) \leq i + k^{-1} \leq i + 1, \; k = 1, \ldots .$$

If $v^o_k(t) = f_o(t, x_k(t), u_k(t))$, $t_{1k} \leq t \leq t_{2k}$, then we define the functions $v^-_k(t)$, $v^+_k(t)$ as follows :

$$v^-_k(t) = - M_o , \quad v^+_k(t) = v^o_k(t) + M_o = f_o(t, x_k(t), u_k(t)) + M_o .$$

Then $v^-_k(t) \leq 0$, $v^+_k(t) \geq 0$ a.e. in $[t_{1k}, t_{2k}]$, and we define

$$y_k^-(t) = \int_{t_{1k}}^{t} v_k^-(t)dt, \quad y_k^+(t) = \int_{t_{1k}}^{t} v_k^+(t)\,dt, \qquad t_{1k} \leq t \leq t_{2k}, k=1,2,\dots .$$

Since $v_k^-(t) = -M_o$, we have $y_k^-(t) = -M_o(t-t_{1k}) \leq 0$, and the functions $y_k^-(t)$ are monotone nonincreasing and uniformly Lipschitzian with constant M_o. On the other hand, the functions $y_k^+(t)$ are nonnegative, monotone nondecreasing,, and uniformly bounded since

$$0 \leq y_k^+(t_{2k}) = (y_k^+(t_{2k}) + y_k^-(t_{2k})) - y_k^-(t_{2k}) = x_k^o(t_{2k}) - y_k^-(t_{2k})$$

$$\leq i + 1 + M_o(t_{2k} - t_{1k}) \leq DM_o + |i| + 1 .$$

By Ascoli's theorem we first extract a sequence for which $(x_k(t), y_k^-(t))$, $t_{1k} \leq t \leq t_{2k}$, converges in the ρ- metric toward continuous vector function $(x(t)\ \ Y^-(t))$, $t_1 \leq t \leq t_2$. Here $x(t)$ is AC because of the equiabsolute continuity of the vector functions $x_k(t)$, and $y^-(t) = -M_o(t-t_1)$. $Y^-(t_1) = 0$. Then we apply Helly's theorem to the sequence $y_k^+(t)$ and we perform a successive extraction so that the corresponding sequence of the $y_k^+(t)$ converges for every $t_1 < t < t_2$ toward a function $Y_o^+(t)$, $t_1 < t < t_2$, which is nonnegative monotone, nondecreasing, but not necessarily continuous. We define $Y_o^+(t)$ at t_1 by taking $Y_o^+(t_1) = 0$, and at t_2 by continuity at t_2, because of its monotoneity. Thus $0 \leq Y_o^+(t) \leq DM_o + |i| + 1$, $t_1 \leq t \leq t_2$.

Finally, $Y_o^+(t)$ admits of a unique decomposition $Y_o^+(t) = Y^+(t) + Z(t)$, $t_1 \leq t \leq t_2$, with $Y^+(t_1) = 0$, where both $Y^+(t)$, $Z(t)$ are nonnegative monotone nondecreasing, where $Y^+(t)$ is AC, and $Z'(t) = 0$ a.e. in $[t_1, t_2]$. Finally, if $Y(t) = Y^-(t) + Y^+(t)$, we see that $x_k^o(t)$, $t_{1k} \leq t \leq t_{2k}$, $k = 1, 2, \dots$, converges for all $t_1 < t < t_2$, toward $x^o(t) = Y(t) + Z(t)$, where $Y(t)$ is a (scalar) AC function, $DM_o \leq Y(t) \leq DM_o + |i| + 1$, $Y(t_1)=0$.

Let us prove that $Y(t_2) \leq i$. For the subsequence $[k]$ we have extracted last, we have $t_{2k} \to t_2$, $x_k^o(t_{2k}) \to i$, $x_k^o(t_{2k}) = y_k^-(t_{2k}) + y_k^+(t_{2k})$. If $\bar{t}_2$ is any point, $t_1 < \bar{t}_2 < t_2$, $\bar{t}_2$ as close as we want to t_2, then $\bar{t}_2 < t_{2k}$ for all k sufficiently large (of the extracted sequence), since $t_{2k} \to t_2$. We can assume k so large that $\bar{t}_2 < t_{2k}$, $|\bar{t}_2 - t_{2k}| < 2|\bar{t}_2 - t_2|$. Then

$$|y_k^-(\bar{t}_2) - y_k^-(t_{2k})| = M_o|\bar{t}_2 - t_{2k}| \leq 2M_o|\bar{t}_2 - t_2| .$$

Since $y_k^+(t)$ is nondecreasing, we have $y_k^+(\bar{t}_2) < y_k^+(t_{2k})$, and finally

$$\begin{aligned} y_k^-(\bar{t}_2) + y_k^+(\bar{t}_2) &\leq y_k^-(\bar{t}_2) + y_k^+(t_{2k}) \\ &\leq y_k^-(t_{2k}) + y_k^+(t_{2k}) + |y_k^-(\bar{t}_2) - y_k^-(t_{2k})| \\ &\leq x_k^o(t_{2k}) + 2M_o|\bar{t}_2 - t_2| , \end{aligned}$$

where $x^o(t_{2k}) \to i$ as $k \to +\infty$, and $x^o(t_{2k}) < i + k^{-1}$. Hence

$$y_k^-(\bar{t}_2) + y_k^+(\bar{t}_2) < i + 2M_o|\bar{t}_2 - t_2| + k^{-1} .$$

As $k \to +\infty$ (along the extracted sequence), we have

$$Y^-(\bar{t}_2) + Y_o^+(\bar{t}_2) \leq i + 2M_o|\bar{t}_2 - t_2| ,$$

or

$$Y^-(\bar{t}_2) + Y^+(\bar{t}_2) + Z(\bar{t}_2) \leq i + 2M_o|\bar{t}_2 - t_2| ,$$

where the third term in the first member is ≥ 0. Thus

$$Y(\bar{t}_2) = Y^-(\bar{t}_2) + Y^+(\bar{t}_2) \leq i + 2M_o|\bar{t}_2 - t_2| .$$

As $\bar{t}_2 \to t_2 - 0$, we obtain $Y(t_2) \le i$, since Y is continuous at t_2.

We will apply below closure theorem II to an auxiliary problem which we shall now define. Let $\tilde{u} = (u^o, u) = (u^o, u^1, \ldots, u^m)$, let $\tilde{U}(t, x)$ be the set of all $\tilde{u} \in E_{m+1}$ with $u = (u^1, \ldots, u^m) \in U(t, x)$, $u^o \ge f_o(t, x, u)$, let $\tilde{x} = (x^o, x) = (x^o, x^1, \ldots, x^n)$, let $\tilde{\tilde{f}} = \tilde{\tilde{f}}(t, x, u) = (\tilde{f}_o, f) = (\tilde{f}_o, f_1, \ldots, f_n, f_n)$ with $\tilde{f}_o = u^o$. Thus, $\tilde{\tilde{f}}$ depends only on $t, x, \tilde{u}$ (instead of $t, \tilde{x}, \tilde{u}$), and U depends only on t, x (instead of $t, \tilde{x}$). Finally, we consider the differential system

$$d\tilde{x}/dt = \tilde{\tilde{f}}(t, x, u),$$

or

$$dx^o/dt = u^o(t), \quad dx^i/dt = f_i(t, x, u), \quad i = 1, \ldots, n,$$

with the constraints

$$\tilde{u}(t) \in \tilde{U}(t, x(t)),$$

or

$$u^o(t) \ge f_o(t, x(t), u(t)), \quad u(t) \in U(t, x, (t)),$$

a.e. in $[t_1, t_2]$, with moreover $x^o(t_1) = 0$, $(t, x(t)) \in A$, and $[x, u] \in \Omega$. We have here the situation discussed in closure theorem II, where $\tilde{x}$ replaces x, x replaces y, x^o replaces z, n + 1 replaces n, n replaces s, hence $(n + 1) - n = 1$ replaces $n - s$. For the new auxiliary problem the cost functional is

$$J[x, u] = \int_{t_1}^{t_2} \tilde{f}_o \, dt = \int_{t_1}^{t_2} u^o(t)dt = x^o(t_2).$$

Note that the set $\tilde{\tilde{Q}}(t,x) = \tilde{\tilde{f}}(t,x,\tilde{U}(t,x))$ of the new problem is the set of all $\tilde{z} = (z^o, z) \in E_{n+1}$ such that $z^o = u^o$, since $\tilde{f}_o = u^o$, $z = f(t,x,u)$, $u^o \geq f_o(t,x,u)$, $u \in U(t,x)$. Thus, the sets $\tilde{\tilde{U}}$, $\tilde{\tilde{Q}}$ for this auxiliary problem are the sets $\tilde{U}$, $\tilde{\tilde{Q}}$ considered before.

We consider now the sequence of trajectories $\tilde{x}_k(t) = [x_k^o(t), x_k(t)]$, $t_{1k} \leq t \leq t_{2k}$, for the problem $J[\tilde{x}, \tilde{u}]$ corresponding to the control function $\tilde{u}_k(t) = [u_k^o(t), u_k(t)]$, with $u_k^o(t) = f_o(t, x_k(t), u_k(t))$, $u_k(t) \in U(t, x_k(t))$, and hence $\tilde{u}_k(t) \in \tilde{U}(t, x_k(t))$, $t_{1k} \leq t \leq t_{2k}$, $k = 1, 2, \ldots$. The sequence $[x_k(t)]$ converges in the metric ρ toward the AC vector function $x(t)$, while $x_k^o(t) \to x^o(t)$ as $k \to +\infty$ for all $t \in (t_1, t_2)$, and $x^o(t) = Y(t) + Z(t)$, where $Y(t)$ in AC in $[t_1, t_2]$ and $Z'(t) = 0$ a.e. in $[t_1, t_2]$.

By closure theorem II we conclude that $X(t) = [Y(t), x(t)]$ is a trajectory for the problem . In other words, there is a control function $\tilde{u}(t)$, $t_1 \leq t \leq t_2$, $\tilde{u}(t) = (u^o(t), u(t))$, with

$$dY/dt = u^o(t) \geq f_o(t, x(t), u(t)), \quad u(t) \in U(t, x(t)),$$

$$(24) \qquad dx/dt = f(t, x(t), u(t)),$$

a.e. in $[t_1, t_2]$, and

$$(25) \qquad i \geq Y(t_2) = J[\tilde{x}, \tilde{u}] = \int_{t_1}^{t_2} u^o(t)\, dt .$$

First of all $[x(t), u(t)]$ is admissible for the original problem, and hence belongs to Ω, since by hypothesis Ω is complete. From this remark, and relations (24) and (25) we deduce

$$i \leq I[x,u] = \int_{t_1}^{t_2} f_o(t, x(t), u(t))dt \leq \int_{t_1}^{t_2} u^o(t)dt < i,$$

and hence all $\leq$ signs can be replaced by = signs, $= u^o(t) = f_o(t, x(t), u(t))$ a.e. in $[t_1, t_2]$, and $I[x,u] = i$. This proves that i is attained in Ω. Thus Existence Theorem I is proved in the case A is compact.

Let us assume now that A is not compact but closed, that A is contained in a slab $[t_o \leq t \leq T, -\infty < x^i < +\infty, i = 1, \dots, n, t_o$ T finite$]$, and that the additional hypotheses (a) and (b) hold. If $Z(t)$ denotes the scalar function $Z(t) = |x(t)|^2 + 1$, then condition $x^1 f_1 + \dots + x^n f_n \leq N(|x|^2 + 1)$ implies $Z' \leq 2NZ$, and hence, by integration from t^* to t, also

$$1 \leq Z(t) \leq Z(t^*) \exp 2N|t - t^*|.$$

Since $[t^*, x(t^*)] \in P$ where P is a compact subset of A, then there is a constant N_o such that $|x| \leq N_o$ for every $x \in P$, hence $1 \leq Z(t^*) \leq N_o^2 + 1$, and $1 \leq Z(t) \leq (N_o^2 + 1) \exp 2N(T - t_o)$. Thus, for $t_o \leq t \leq T$ $Z(t)$ remains bounded, and hence $|x(t)| \leq D$ for some constant D. We can now restrict ourselves to the consideration of the compact part A_o of all points (t, x) of A with $t_o \leq t \leq T$, $|x| \leq D$.

Thus, theorem I is proved for A closed and contained in a slab as above, and under the additional hypotheses (a), (b).

Let us assume that A is not compact, nor contained in any slab as above, but closed, and that hypotheses (a), (b), (c) hold. First, let us take an arbitrary element $\bar{x}(t)$, $\bar{u}(t)$ of Ω and let

$j = I[\bar{x}, \bar{u}]$. Then we consider a bounded interval (a, b) of the t-axis containing the entire projection P_o of P on the t-axis , as well as the interval $[-R, R]$. Now let $\ell = \mu^{-1}[\,|j| + 1 + (b-a) M_o]$, where $f_o(t, x, u) \geq -M_o$ for all $(t, x, u) \in M$ with $(t, x) \in A_o$, the compact part of A with $t \in [a, b]$ and $|x| \leq D$, D as determined previously, and $M_o \geq 0$.

Let $[a', b']$ denote the interval $[a-\ell, b+\ell]$. Then for any admissible pair (if any) $x(t)$, $u(t)$, $t_1 \leq t \leq t_2$, of the class Ω , whose interval $[t_1, t_2]$ is not contained in $[a', b']$ there is at least one point $t^* \in [t_1, t_2]$ with $((t^*, x(t^*)) \in P$, $a < t^* < b$, and a point $\bar{t} \in [t_1, t_2]$ outside $[a', b']$. Hence $[t_1, t_2]$ contains at least one subinterval, say E, outside $[a, b]$, of measure $\geq \ell$. Then $I[x, u] \geq \geq \ell\mu - (b-a) M_o = |j| + 1 \geq i + 1$. Obviously, we may disregard all pairs $x(t)$, $u(t)$, $t_1 \leq t \leq t_2$, whose interval $[t_1, t_2]$ is not contained in $[a', b']$. In other words, we can limit ourselves to the closed part A' of all $(t, x) \in A$ with $a' \leq t \leq b'$. We are now in the situation above, and theorem I is proved for any closed set A under the additional hypotheses (a), (b) , (c) . Finally, we have to show that condition (a) can be replaced by condition : (a') There are numbers $C, D > 0$ such that $f_o(t, x, u) \geq C|f(t, x, u)|$ for all $(t, x, u) \in M$ with $|x| \geq D$. It is enough to prove theorem I under the hypotheses that A is closed and contained in a slab $t_o \leq t \leq T$, t_o, T as above, and hypotheses (a') and (b) . First let us take D so large that the projection P^* of P on the x-space is completely in the interior of the solid sphere $|x| \leq D$, and also so large that $D \geq T - t_o$. Let $\bar{x}(t)$, $\bar{u}(t)$ be any arbitrary admissible pair contained in Ω , and let j denote the corresponding value of the cost functional. Let $L = C^{-1}\{DM_o + |j| + 1]$, and let us take $D_o = D + L$. If any admissible pair $x(t), u(t), t_1 \leq t \leq t_2$, of Ω possesses a point $(t_o, x(t_o))$ with $|x(t_o)| \geq D_o$, then $x(t)$ possesses also a point $(t^*, x(t^*)) \in P$, with $|x(t^*)| < D$. Thus, there is at least a subarc $\Gamma : x = x(t)$, $t' \leq t \leq t''$, of $x(t)$ along which $|x(t)| \geq D$ and $|x(t)|$ passes from the value D to the value $D_o = D + L$. Such an arc Γ has a length $\geq L$. If $E = [t_1, t_2] - [t', t'']$, then

$$I[x, u] = \int_{t_1}^{t_2} f_o \, dt = \left(\int_E + \int_{t'}^{t''} \right) f_o \, dt \geq -DM_o + \int_{t'}^{t''} C|f| \, dt$$

$$= -DM_o + C \int_{t'}^{t''} |dx/dt| \, dt = -DM_o + CL = |j| + 1 \geq i+1.$$

where $-M_o$ again is a lower bound for $f_o(t, x, u)$ for each $(t, x, u) \in M$ with $(t, x) \in A_o$, the compact part of A with $t_o \leq t \leq T$, $|x| \leq D$. As before we can restrict ourselves to the compact part A_o of all points (t, x) of A with $t_o \leq t \leq T$, $|x| \leq D$. The case where A is closed, A is not contained in any slab as above, but conditions (a'), (b), (c) hold can be treated as before. The case where A is not compact and conditions (e), (f), (g) hold, also can be treated as before. Theorem I is thereby completely proved.

9. Another Existence Theorem for Lagrange Problems with Unilateral Constraints . Existence Theorem II (L. Cesari [1 e]).

Let A be a compact subset of the tx-space $E_1 \times E_n$, and, for every $(t, x) \in A$, let $U(t, x)$ be a closed subset of the u-space E_m. Let $\tilde{f}(t, x, u) = (f_o, f_1, \ldots, f_n) = (f_o, f)$ be a continuous vector function on the set M of all (t, x, u) with $(t, x) \in A$, $u \in U(t, x)$, Assume that for every $(t, x) \in A$, the set

$$\tilde{\tilde{Q}}(t,x) = [\tilde{z} = (z^0, z) \mid z^0 \geq f_0(t,x,u),\ z = f(t,x,u),\ u \in U(t,x)] \subset E_{n+1}$$

is convex, and that $U(t,x)$ satisfies property (U) and $\tilde{\tilde{Q}}(t,x)$ satisfies property (Q) in A. Let $\varphi(t)$ be a given function which is L-integrable in any finite interval such that $f_0(t,x,u) \geq \varphi(t)$ for all $(t,x,u) \in M$. Let Ω be a nonempty complete class of admissible pairs $x(t)$, $u(t)$ such that

$$(26) \qquad \int_{t_1}^{t_2} |dx^i/dt|^p \, dt \leq N_i, \quad i = 1, \dots, n$$

for some constants $N_i \geq 0$, $p > 1$, Then the cost functional $I[x, u]$ has an absolute minimum in Ω.

If A is not compact, but closed and contained in a slab $[t_0 \leq t \leq T, -\infty < x^i < +\infty, i = 1, \dots, n, t_0, T \text{ finite}]$, then Existence Theorem II still holds under the additional hypothesis (b) after theorem I. If A is not compact, nor contained in any slab as above, but A is closed then Existence Theorem II still holds under the additional hypotheses (b) and (c^+) : $f_0(t,x,u) \geq \varphi(t)$ for all $(t,x,u) \in M$ where $\varphi(t)$ is a given function which is L-integrable in any finite interval and $\int_0^{+\infty} \varphi(t)\,dt = +\infty$, $\int_{-\infty}^0 \varphi(t)\,dt = +\infty$. Finally, if for some $i = 1, \dots, n$, and any $N > 0$, there is some $N_i > 0$ such that $(x,u) \in \Omega$, $I[x, u] \leq N$ implies $\int_{t_1}^{t_2} |dx^i/dt|^p dt \leq N_i$ then the corresponding requirements (26) can be disregarded.

<u>Proof of Existence Theorem II.</u> We suppose A compact, hence necessarily contained in a slab $[t_o \le t \le T,\ t_o,\ T$ finite, $-\infty < x^i < +\infty,\ i = 1, \dots, n]$, and then $I[x, u] = \int_{t_1}^{t_2} f_o\, dt \ge \int_{t_o}^{T} \varphi(t)\, dt$. This proves that the infimum i of $I[x, u]$ in Ω is necessarily finite. Let $u_k(t)$, $x_k(t)$, $t_{1k} \le t \le t_{2k}$, $k = 1, 2, \dots$, be a sequence of admissible pairs all in with $I[u_k, x_k] \to i$. We may assume

$$(27) \qquad i \le I[x_k, u_k] = \int_{t_{1k}}^{t_{2k}} f_o(t, x_k(t), u_k(t))\, dt \le i + 1/k \le i + 1 .$$

Then

$$(28) \qquad \int_{t_{1k}}^{t_{2k}} |dx_k^i/dt|^p dt \le N_i, \qquad i = 1, \dots, n,\ k = 1, 2, \dots .$$

By the weak compactness properties of L_p we conclude that there is some AC vector function $x(t) = (x^1, \dots, x^n)$, $t_1 \le t \le t_2$, such that $t_{1k} \to t_1$, $t_{2k} \to t_2$, $dx_k^i/dt \to dx^i/dt$ weakly in L_p, $x_k(t) \to x(t)$ in the ρ-metric. The proof is now exactly the same as for Existence Theorem I.

If A is <u>not</u> compact, but closed and contained in a slab as above, and condition (b) holds, then for every admissible pair x(t), u(t) of Ω we have

$$|x(t) - x(t^*)| = \left|\int_{t^*}^{t} (dx/dt)dt\right| \left[\le \int_{t^*}^{t} dt\right]^{1/q} \left[\int_{t^*}^{t} |dx/dt|^p dt\right]^{1/p}$$

$$\le |t - t^*|^{1/q} (N_1 + \dots + N_n), \quad q^{-1} + p^{-1} = 1.$$

where $(t^*, x(t^*))$ belongs to a fixed compact subset P of A. Then $|x(t^*)| \le N'$, $|t - t^*| \le T - t_o$, and $|x(t)| \le N''$ for some constant N', $N'' > 0$. Thus, we can limit ourselves to the compact part A_o of all points (t,x) of A with $t_o \leq t \leq T$, $|x| \leq N''$. If A is not compact, nor contained in any slab as above, but A is closed and conditions (b), (c) hold, then we can use the same argument as for Existence Theorem I.

Finally, we see that assumption (26) has been used only in (28) for a minimizing sequence x_k, u_k. Since for a minimizing sequence we see already in (27) that $I[x_k, u_k] \leq i+1$, it is obvious that any relation (26) which is a consequence of a relation of the form $I \geq M$ need not be required among the assumptions of theorem II. Existence Theorem II is thereby proved.

10. <u>A few corollaries.</u>

<u>Corollary 1.</u> (an existence theorem for Pontryagin's problem). Let A be any compact subset of the tx-space $E_1 \times E_n$, and for every $(t,x) \in A$ let $U(t,x)$ be a compact subset of the u-space E_m. Let M be the set of all (t,x,u) with $(t,x) \in A$, $u \in U(t,x)$, and let $\tilde{f}(t,x,u) = (f_o, f) = (f_o, f_1, \dots, f_n)$ be a continuous vector function on M. Let $U(t,x)$ be an upper semicontinuous function of (t,x) in A, and for every $(t,x) \in A$ let

$$\tilde{\tilde{Q}}(t,x) = \left\{\tilde{z} = (z^o, z) \,|\, z^o \geqq f_o(t,x,u),\ z = f(t,x,u),\ u \in U(t,x)\right\} \subset E_{n+1}$$

be a convex subset of E_{n+1}. Let B be a closed subset of the $t_1x_1t_2x_2$-space E_{2n+2}. Then the cost functional $I[x,u]$ has an absolute minimum in any nonempty complete class Ω of admissible pairs $u(t)$, $x(t)$.

If A is not compact but closed, the statement holds under the same additional hypotheses listed under Theorem I

<u>Proof.</u> Here M is a compact set, and hence $|f_o(t,x,u)| \leq M_o$ for all $(t,x,u) \in M$ and some constant M_o. If we denote here by $\tilde{\tilde{Q}}_o(t,x)$ the intersection of the set $\tilde{\tilde{Q}}$ of the text with the set $[\tilde{z} = (z^o, z) \mid z^o \leq M_o]$, then this set $\tilde{\tilde{Q}}_o(t,x)$ can be used instead of $\tilde{\tilde{Q}}$ in the reasoning of Theorem I. On the other hand, $\tilde{\tilde{Q}}_o(t,x)$ is now a compact convex subset of E_{n+1} for every $(t,x) \in A$, and $U(t,x)$ has property (U), and $\tilde{\tilde{Q}}_o(t,x)$ is upper semicontinuous and has property (Q). Also $|f(t,x,u)| \leq M_1$ for all $(t,x,u) \in M$ and some constant M_1, and if we take $C = M_1$, $D = 0$, then $|f(t,x,u)| \leq M_1 = C + D|u|$ for all $(t,x,u) \in M$. Also, if D_1 is a constant such that $|u| \leq D_1$ for all (t,x,u) of the compact set M, and we take $\varphi(\zeta) = -G + \zeta^2$, $G = M_o + D_1^2$, then

$$f_o(t,x,u) \geq -M_o \geq -M_o + (|u|^2 - D_1^2) = \varphi(|u|).$$

All conditions of Existence Theorem I are satisfied, and Corollary 1 is thereby proved

<u>Corollary 2</u> (Filippov's existence theorem for Pontryagin's problems).

Same as corollary 1 with the hypothesis of convexity of $\tilde{Q}(t,x)$ replaced by the hypothesis of convexity of $\tilde{Q}(t,x)$.

Indeed, if the set $Q(t,x)$ in no. 7 is convex for every $(t,x) \in A$, then certainly $Q(t,x)$ and $Q_o(t,x)$ are convex.

Remark. The generality of theorem I, or of corollary 1, can be seen best in the case of free problems. Indeed, then for the convexity of $\tilde{Q}(t,x)$ as requested in Filippov's statement we need $f_o(t,x,u)$ to be linear in u for every $(t,x) \in A$, while for the convexity of $\tilde{\tilde{Q}}(t,x)$ as requested in Theorem I, or in corollary 1, we need only $f_o(t,x,u)$ to be convex in u for every $(t,x) \in A$, as already mentioned in no. 7.

Corollary 3 (the Nagumo-Tonelli existence theorem for free problems).

If A is a compact subset of the tx-space $E_1 \times E_n$, if $f_o(t,x,u)$ is a continuous scalar function on the set $M = A \times E^n$, if for every $(t,x) \in A$, $f_o(t,x,u)$ is convex as a function of u in E, if there is a continuous scalar function $\varphi(\zeta)$, $0 \le \zeta < +\infty$, with $\varphi(\zeta)/\zeta \to +\infty$ as $\zeta \to +\infty$, such that $f_o(t,x,u) \ge \varphi(|u|)$ for all $(t,x,u) \in M$, then the cost functional.

$$I[x] = \int_{t_1}^{t_2} f_o(t, x(t), x'(t))\, dt$$

has an absolute minimum in any nonempty complete class Ω of absolutely continuous vector functions $x(t)$, $t_1 \le t \le t_2$, for which $f_o(t,x(t),x'(t))$ is L-integrable in $[t_1, t_2]$.

If A is not compact, but closed and contained in a slab $[t_o \le t \le T,\ x \in E_n]$, t_o, T finite, then the statement still holds under the additional hypotheses $(\mathcal{T}_1)$ $f_o \ge C|u|$ for all $(t,x,u) \in M$ with $|x| \ge D$ and convenient constants $C > 0$, $D \ge 0$; $(\mathcal{T}_2)$ every trajectory $x(t)$ of Ω possesses at least one point $(t^*, x(t^*))$ on a given compact subset P of A. If A is not compact, nor contained in a slab as above, but A is closed, then the statement still holds under the additional hypotheses $(\mathcal{T}_1)$, $(\mathcal{T}_2)$, and $(\mathcal{T}_3)$ $f_o(t, x, u) \geq \mu > 0$ for all $(t, x, u) \in M$ with $|t| \geq R$, and convenient constants $\mu > 0$ and $R > 0$.

Proof. The free problem under consideration can be written as a Lagrange problem with $m = n$, $f_i = u_i$, $i = 1, \ldots, n$, $U(t, x) = E_m = E_n$, so that the differential system reduces to $dx/dt = u$, and the cost functional is

$$I[x, u] = \int_{t_1}^{t_2} f_o(t, x(t), u(t))\, dt = \int_{t_1}^{t_2} f_o(t, x(t), x'(t))dt.$$

First assume A to be compact. Then the set $\tilde{Q}(t, x)$ reduces here to the set of all $\tilde{z} = (z^o, z) \in E_{n+1}$ with $z^o \geq f_o(t, x, z)$, $z \in E_n$, where f_o is convex in z, and satisfies the growth conditon $f_o \geq \varphi(|u|)$ with $\varphi(\zeta)/\zeta \to +\infty$ as $\zeta \to +\infty$. Thus $\tilde{\tilde{Q}}_o(t, x)$ is convex subset of E_{n+1} for every $(t, x) \in A$. By a remark in [1e], $\tilde{\tilde{Q}}_o$ satisfies condition (Q) in A. Thus, all hypotheses of theorem I are satisfied. If A is closed but contained in a slab as above then the condition (a) of theorem I reduces to $u x = C(x^2 + 1)$ which cannot be satisfied since we have no bound on u. On the other hand, the condition (a') $f_o \geq C\,|f|$ for all $(t, x, u) \in M$ with $|x| \geq D$ and some constants $C > o$, $D > o$, reduces here to requirement $(\mathcal{T}_1)$ and condition (b) to requirement $(\mathcal{T}_2)$. Finally; if A is not compact, nor contained in a slab as above, but A is closed, then requirement (c) of Existence Theorem I reduces to requirement $(\mathcal{T}_3)$. All conditions of Existence Theorem I are satisfied and the cost functional has an absolute minimum in Ω.

II. Examples

(a) Let us consider the (free) problem

$$I(x) = \int_{t_1}^{t_2} (1 + |x'|^2) \, dt = \text{minimum},$$

with $x = (x^1, \dots, x^n)$, in the class Ω of all absolutely continuous functions $x(t) = (x^1, \dots, x^n)$, $0 \le t \le t_2$, whose graph $(t, x(t))$ joins the point $(t_1 = 0,\ x(t_1) = (0, \dots, 0))$ to a nonempty closed set B of the half-space $t_2 \geq 0$, $x \in E_n$. This problem can be written as a Lagrange problem :

$$J[x, u] = \int_{t_1}^{t_2} (1 + |u(t)|^2) \, dt = \text{minimum} \ ,$$

$$dx^i/dt = u^i, \quad i = 1, \dots, n ,$$

where $x(t) = (x^1, \dots, x^n)$, $u(t) = (u^1, \dots, u^n)$, $m = n$, $f_o = 1 + |u|^2$, $f_i = u^i$, $i = 1, \dots, n$, and the control space $U(t, x)$ is fixed and coincides with the whole space E_n. Here $\tilde{\tilde{Q}}(t, x) = [(x, u) \mid z > 1 + |u|^2,\ u \in E_n]$ is a fixed and convex subset of E_{n+1}. The conditions of Existence Theorem I, as well as those of corollary 3, are satisfied with $\phi(\zeta) = \zeta^2$, $0 \le \zeta < +\infty$, and A the half-space $A = [(t, x) \mid t \geq 0,\ x \in E_n] \subset E_{n+1}$. Thus the problem above has an optimal solution.

(b) As an example of application of corollary 1 we may consider the Pontryagin problem with $m = n = 2$

$$I = \int_{t_1}^{t_2} (x^2 + y^2 + u^2 + v^2 + 1) dt = \text{minimum},$$

$$U(x, y) = \left\{ -1 \leqq u \leqq 1,\ -1 \leqq v \leqq 1 \right\},$$

$$dx/dt = u, \qquad dy/dt = v$$

$$t_1 = 0, \quad x(0) = y(0) = 0, x(t_2) = 1, \quad t_2,\ y(t_2) \text{ undetermined}.$$

Here A is the halfspace (t, x, y) with $t \geqq 0$, $(x, y) \in E_2$. Also, B is the closed set $B = \{(t_1, x_1, y_1, t_2, x_2, y_2) \mid t_1 = 0,\ x_1 = 0,\ y_1 = 0,\ x_2 = 1\}$, that is, a 2-plane in E_6, and B_1 is the single point $(0, 0, 0)$ of E_3, certainly a compact set, $P = B_1$, and condition (b) is satisfied. The set

$$\tilde{Q}(x,y) = \left\{ (z, u, v) \mid z = x^2 + y^2 + u^2 + v^2 + 1,\ -1 \leqq u \leqq 1,\ -1 \leqq v \leqq 1 \right\}$$

is not convex, and Filippov's theorem does not apply. Instead

$$\tilde{\tilde{Q}}(x, t) = \left\{ (z, u, v) \mid z \geqq x^2 + y^2 + u^2 + v^2 + 1,\ -1 \leqq u \leqq 1,\ -1 \leqq v \leqq 1 \right\}$$

is certainly convex. Finally, by the usual relation $|x| \leqq 2^{-1}(1 + x^2)$, x scalar. we deduce

$$xu + yv \leqq |x| + |y| \leqq 1 + x^2 + y^2,$$

and condition (a) is satisfied. Condition (c) is satisfied with $G = \mu = 1$. Thus the problem possesses an optimal solution by force of corollary 1.

(c) As an example of application of corollary 2 we may consider the Pontryagin problem with $m = n = 2$, .

$$I = \int_{t_1}^{t_2} (x^2 + y^2 + 1)dt = \text{minimum} ,$$

$$U = U(t, x, y) = [-1 \leq u \leq 1, \ -1 \leq v \leq 1] ,$$

$$dx/dt = u , \qquad dy/dt = v,$$

$$t_1 = 0, \ x(0) = y(0) = 0, \ x(t_2) = 1, \ t_2, y(t_2) \text{ undetermined.}$$

(d) Another example of application of corollary 2 is the problem of minimum transfer time ($f_o = 1$),

$$I = \int_{t_1}^{t_2} dt = t_2 - t_1 ,$$

$$U = U(t, x, y) = [- 1 \leq u \leq 1, -1 \leq v \leq 1]$$

$$dx/dt = u , \ dy/dt = v,$$

$$t_1 = 0, \ x(0) = y(0) = 1, \ x(t_2) = 1, \ (t_2, \ y(y(t_2)) \text{ undetermined.}$$

(e) The free problem

$$I[x] = \int_o^1 tx'^2 dt = \text{minimum}, \quad x(0) = 1, \ x(1) = 0,$$

is known to have no optimal solution . Indeed, if Ω is the class of all AC scalar functions $x(t)$, $0 \leq t \leq 1$, satisfying the boundary conditions above and for which tx'^2 is L-integrable in $[0, 1]$, then $I[x] \geq 0$, and the infimum i of $I[x]$ in Ω is $i \geq 0$. On the other hand, if we take $x_k(t) = 1$ for $0 \leq t \leq k^{-1}$, $x_k(t) = -\log t/\log k$ for $k^{-1} \leq t \leq 1$, $k = 2, 3, \ldots$, then by computation we obtain $I[x_k] = (\log k)^{-1}$, and hence $I[x_k] \to o$ as $k \to +\infty$. Thus $i \leq 0$, and finally $i = 0$. Now $I[x]$ cannot take the value zero in Ω, since $I[x] = 0$ implies $tx'^2(t) = 0$, or $x'(t) = 0$, a.e. in $[0, 1]$, hence $x(t) = \text{constant}$ in $[0, 1]$, in contradiction with the boundary conditions. The simple free problem above has no minimum.

The same problem above can be written as a Lagrange problem with $m = n = 1$ in the form

$$J_1[x, u] = \int_0^1 tu^2 dt = \text{minimum}, x(0) = 1, x(1) = 0,$$
$$dx/dt = u, \quad u \in E_1,$$

as well as in the form

$$J_2[x, u] = \int_0^1 t^3 u^2 dt = \text{minimum}, \; x(0) = 1, \; x(1) = 0, \; dx/dt = tu,$$
$$u \in E_1.$$

The relative sets $\tilde{\tilde{Q}}(t, x,)$ are here subsets of the z^0, z-plane E_2. For neither problem J_1 nor J_2 the integrand function f_0 satisfies the growth condition of Existence Theorem I. For the problem J_2 the sets $\tilde{\tilde{Q}}$ do not satisfy condition (Q).

The same free problem with an additional constraint

$$\int_0^1 x'^2 dt \leq N_0,$$

where $N_0 \geq 1$ any constant, has an optimal solution by force of Existence Theorem II and subsequent remark. The optimal solution will depend on N_0. Note that $N_0 \geq 1$ assures that the class Ω relative to the problem is not empty. Indeed for $x(t) = 1 - t$, we have $\int_0^1 x'^2 \, dt = 1$.

L. Cesari

12 Further existence theorems for Lagrange problems.

Let us assume for a moment A compact. The condition of theorem 1

(α) $f_o(t, xu) \geq \phi(|u|)$ for all $(t, x, u) \in M$,
where $\phi(\zeta)$, $0 \leq \xi < +\infty$, is a continuous function satisfying $\phi(\zeta)/\zeta \to +\infty$ as $\zeta \to +\infty$; $|f(t, x, u)| \leq A+B|u|$ for all $(t, x, u) \in M$ and some constants A, B ≥ 0;

is usually called a "growth conditon". This condition (α) obviously implies

(β) $\lim\limits_{\substack{|u| \to +\infty \\ |u| \to \infty}} |u|^{-1} f_o(t, x, u) = +\infty$ for every $(t, x) \in A$ where
with $u \in U(t, x)$, $(t, x) \in A$: $|f(t, x, u)| \leq A + B|u|$ for all $(t, x, u) \in M$ and some constants $A, B \geq o$;

is also a "growth condition", and examples show that in general (β) is a condition weaker than (α). Nevertheless, for free problems, when f_o is assumed to be convex in u and $U(t, x) \equiv E_n$, it has been proved that (β) is equivalent to (α) (L. Tonelli [16a] for f_o of class C^1; L. Turner [17] for f_o only continuous as assumed here).

A condition slightly stronger than (β) has been taken into consideration, say (γ), and we state it here again for A compact:

(γ) $\lim\limits_{|u| \to +\infty} |u|^{-1} f_o(t, x, u) = +\infty$ uniformly for $(t, x) \in A$ when $|u| \to +\infty$

with $u \in U(t,x)$; $|f(t,x,u)| \leq A + B|u|$ for all $(t,x,u) \in M$ and some constants $A, B \geq 0$.

For A not compact but closed condition (γ) can be replaced by the modified condition $(\gamma)_m$ defined by

$(\gamma)_m$ $\lim_{|u| \to +\infty} |u|^{-1} f_o(t,x,u) = +\infty$ uniformly for (t,x) in any compact part A_o of A, and when $|u| \to +\infty$ with $u \in U(t,x)$; for every compact part A_o of A there are constants $A_o, B_o \geq 0$ such that $|f(t,x,u)| \leq A_o + B_o|u|$ for all $(t,x) \in A_o$, $u \in U(t,x)$.

Existence Theorem III (L. Tonelli [16a], J. R. LaPalm [4]). The same as Existence Theorem I where growth condition (α) is replaced by (γ) if A is compact.

If A is not compact but closed and contained in a slab $[t_o \leq t \leq T,\ x \in E_n]$, t_o, T finite, then theorem III still holds if (a), (b) hold, and (α) is replaced by $(\gamma)_m$. If A is not compact nor contained in a slab as above, but A is closed, then Theorem III still holds if (a), (b), (c) hold, and (α) is replaced by $(\gamma)_m$. Finally, condition (a) can be replaced in any case by condition (d).

The proof is based on modifications of the proof of Theorem I and is omitted. This theorem extends to Lagrange problems a result of Tonelli for free problems ($m=n$, $f=u$, $U=E_n$) [16a, Opere Scelte 3, p. 211]. Indeed, as proved in [17], the convexity of $f_o(t,x,u)$ with respect to u, $U(t,x) = E_n$, and (β) imply $(\gamma)_m$, that is, the uniformity of (β) when (t,x) describes any compact part A_o of A.

For free problems it was shown by Tonelli [16ac] that the growth condition (α) can be dispensed with at the points (t, x) of an exceptional subset E of A provided some additional mild hypotheses is satisfied at the points of E, or E is a suitable "slender" set. This situation repeats for Lagrange problems as we shall state in Theorems IV and V below. We shall need a new condition, say condition (γ^*):

(γ^*) $\lim_{|u| \to +\infty} |u|^{-1} f_o(t, x, u) = +\infty$ uniformly for (t, x) in any compact part A_o of A-E, and where $|u| \to +\infty$ with $u \in U(t, x)$;
For every compact part A_o of A-E, there are constants $A_o, B_o \geq 0$ such that $|f(t, x, u)| \leq A_o + B_o|u|$ for all $(t, x) \in A_o$, $u \in U(t, x)$.

A point $(t_o, x_o) \in A$ is said to possess property (T) provided there is a neighborhood $N_\delta(t_o, x_o)$ in A, two functions $\phi(\zeta)$, $0 < \zeta \leq 1$, $\psi(\zeta)$, $0 < \zeta < +\infty$, and four constants $\ell > 0$, $\alpha > 0$, $\alpha_1 \geq 0$, μ real, with $\phi(\zeta)$ nonnegative, $\phi(0+) = +\infty$, ϕ integrable in (0, 1)), $\psi(\zeta)$ nonnegative, nondecreasing, such that $\zeta\phi(\zeta)\psi(\zeta) \to +\infty$ as $\zeta \to 0+$, and such that $(t, x) \in N_\delta(t_o, x_o)$, $u \in U(t, x)$ implies

$$f_o(t, x, u) \geq |t - t_o|^{\alpha} |u|^{\alpha + \alpha_1} \psi^{\alpha}(|u|) + \mu.$$

In addition, for the same $N(t_o, x_o)$ there are constants $C_o, D_o \geq 0$ such that $|f(t, x, u)| \leq C_o + D_o|u|$ for $(t, x) \in N(t_o, x_o)$, $u \in U(t, x)$.

<u>Existence Theorem IV</u> (L. Tonelli [16a], J.R. LaPalm [4]). The same as Existence Theorem I where a closed exceptional subset E of A is given, condition (T) holds at every point (t, x)

of E, and condition (α) is replaced by (γ^*), if A is compact.

If A is not compact but closed and contained in a slab $[t_o \le t \le T, \ x \in E_n]$, t_o, T finite, then theorem IV still holds if (a), (b), hold, and (α) is replaced by (γ^*). If A is not compact, nor contained in any slab as above, but A is closed, then Theorem IV still holds if (a), (b), (c) hold, and (α) is replaced by (γ^*). Finally, condition (a) can be replaced in any case by condition (d).

The proof is based on modifications of the proof of Theorem I and is omitted. This theorem too extends to Lagrange problems a result of Tonelli for free problems [16a, Opere Scelte 3, p. 213].

If the exceptional set E in suitably "slender", then property (T) at the points (t_o, x_o) of E is not needed. Let E be any subset of A. Then for any subset H of the tx-axis we shall denote by $E^i(H)$ the set of points ξ of the x^i-axis, i = 1, ..., n, defined as follows

$$E^i(H) = \{\xi \mid \text{there exists a point } (t, x^1, \dots, x^n) \in E \text{ with } t \in H, x^i = \xi\}.$$

We shall denote by $\mu^*[E^i(H)]$ the one-dimensional outer measure of $E^i(H)$, and we shall require below that $\mu^*[E^i(H)] = 0$, i = 1, ..., n, for every subset H of the t-axis with measure zero.

For instance, any set E contained in countably many straight lines parallel to the t-axis, and to finitely many (nonparametric) curves $x^i = \varphi_i(t)$, i = 1, ..., N, $t' \le t \le t''$, $\varphi_i(t)$ AC in $[t', t'']$, certainly possesses the property above. The property above.

was proposed and used by L. Turner to extend to free problems in E_n results of Tonelli for free problems in E_1.

Existence Theorem V (L. Tonelli [16a] J.R. LaPalm [4], L. Turner [17]). The same as Theorem I where A is compact and (α) is replaced by (γ^*) provided

(L_1) $\mu^*[E^i(H)] = 0$, $i = 1, \dots, n$, for every subset H of the t-axis of measure zero;

(L_2) for every $(t_o, x_o u_o) \in M$ there are numbers $\delta = \delta(t_o, x_o, u_o) > 0$, $\gamma = \gamma(t_o, x_o, u_o) > 0$, $r = r(t_o, x_o, u_o)$ real, $b_i = b_i(t_o, x_o, u_o)$ real, $i = 1, \dots, n$, such that

$$f_o(t, x, u) \geq r + \sum_{i=1}^{n} b_i f_i(t, x, u) + \gamma f(t, x, u)$$

for all $(t, x) \in N(t_o, x_o)$ $u \in U(t, x)$;

and either condition (L_3) holds, or (L_3') holds:

(L_3) For every compact part A_o of A there is a constant $M_o \geq 0$ such that $f_o(t, x, u) \geq - M_o$ for all $(t, x) \in A_o$ $u \in U(t, x)$;

(L_3') Given $\epsilon > 0$ there are numbers δ, γ, r, b_i as in (L_2) for which (L_2) holds and

$$f_o(t, x, u) < r + \sum_{i=1}^{n} b_i f_i(t, x, u) + \epsilon$$

for all $(t, x) \in N_\delta(t_o, x_o)$, $u \in U(t, x)$, $|u - u_o| \leq \delta$.

If A is not compact but closed and contained in a slab $[t_o \leq t \leq T, \; x \in E_n]$, t_o, T finite, then Theorem V still holds, if (a), (b) hold, and (α) is replaced by (γ^*). If A is not compact nor contained in any slab as above, but A is closed, then Theorem V still holds if (a), (b), (c), hold, and (α) is replaced by (γ^*). Finally, condition (a) can be replaced in any case by condition (d).

The proof is based on modifications of the proof of Theorem I and is mitted . This theorem extends to Lagrange problems a theorem proved by L. Turner for free problems [17] , which in turn ex ends previous results of Tonelli for free problems [16a , Opere Scelte, p. 219] .

L. Cesari

Lecture 5. Existence theorems for generalized solutions of Lagrange problems.

13. Notations. Instead of considering usual cost functional, differential equations and constraints, .

$$I[x, u] = \int_{t_1}^{t_2} f_o(t, x(t), u(t))dt,$$

$$dx/dt = f(t, x(t), u(t)), \qquad f = (f_1, \dots, f_n),$$

$$(t, x(t)) \in A, \quad u(t) \in U(t, x(t)), \quad (t_1, x(t_1), \quad t_2, x(t_2)) \in B,$$

we have to consider the new cost functional, differential equations and constraints,

$$J[x, p, v] = \int_{t_1}^{t_2} g_o(t, x(t), p(t), v(t))dt,$$

$$dx/dt = g(t, x(t), p(t), v(t)), \quad (t_1, x(t_1), t_2, x(t_2)) \in B,$$

$$(t, x(t)) \in A, \quad (p(t), v(t)) \in \Gamma \times V(t, x(t)) .$$

Here $x = (x^1, \dots, x^n)$, $p = (p_1, \dots, p_\gamma)$, $\gamma \geqq n+1$, $v = (u^{(j)}, j = 1, \dots, \gamma)$, $\Gamma = \{p | p_j \geqq 0, p_1 + \dots + p_\gamma = 1\}$, $u^{(j)} \in U(t, x)$, and therefore, the new control variable (p, v) takes on its values in the set $\Gamma \times V(t, x)$, where $V = [U]^\gamma$ is the Cartesian product of U by itself taken γ times. In other words, $v(t) = (u^{(1)}, \dots, u^{(\gamma)})$ represents a finite system of $\gamma \geqq n+1$ ordinary strategies $u^{(1)}, \dots, u^{(\gamma)}$, each $u^{(j)}$ having its values in U, that is, $u^{(j)}(t) \in U(t, x(t)) \subset E_m$, $j = 1, \dots, \gamma$. Thus $v=(u^{(1)}, \dots, u^{(\gamma)})$ is a vector variable whose γ components $u^{(1)}, \dots, u^{(\gamma)}$

are themselves vectors with values in $U(t,x) \subset E_m$, hence $v(t) \in V(t, x(t)) \subset E_{m\gamma}$. As is indicated above $p(t) = (p_1, \dots, p_\gamma)$ represents a probability distribution ; hence p is an element of the simplex $\Gamma \subset E_\gamma$ defined above, and for the new control variable we have $(p(t), v(t)) \in \Gamma \times V(t, x(t)) \subset E_{\gamma+m\gamma}$. As usual, we denote by g and $\tilde{g}$ the two vectors $g = (g_1, \dots, g_n)$, $\tilde{g} = (g_0, g_1, \dots, g_n) = (g_0, g)$, with

$$g_i(t,x,p,v) = \sum_{j=}^{\gamma} p_j f_i(t,x,u^{(j)}), \quad i = 0, 1, \dots, n .$$

As usual, we require that all functions $p_j(t)$, $u^{(j)}(t)$ are measurable, and that x(t) is absolutely continuous.

We say that $[p(t), v(t)]$ is a generalized strategy, that $p(t) = (p_1, \dots, p_\gamma)$ is a probability distribution, and that x(t) is a generalized trajectory. We shall also say for the sake of brevity that $[x(t), p(t), v(t)]$ is a <u>generalized</u> or <u>weak</u> solution.

If we introduce, as usual, the auxiliary variable x^0 with initial value $x^0(t_1) = 0$, and the vector $\tilde{x} = (x^o, x) = (x^0, x^1, \dots, x^n)$ then instead of the system $d\tilde{x}/dt = \tilde{f}$, we shall consider the system

$$d\tilde{x}/dt = \tilde{g}(t, x(t), p(t), v(t)), \qquad \tilde{g} = (t_0, g) = (g_0, g_1, \dots, g_n),$$

and we have

$$J[x,p,v] = x^0(t_2) .$$

Instead of the usual sets $Q(t,x) = f[t,x,U(t,x)] \subset E_n$ and

$$\widetilde{Q}(t,x) = \widetilde{f}(t,x,U(t,x))$$
$$= \{\widetilde{z} = (z^0, z) \mid \widetilde{z} = \widetilde{f}(t,x,u);\ u \in U(t,x)\} \subset E_{n+1} .$$

we shall now consider the sets

$$R(t,x) = g[t,x,\Gamma \times V(t,x)]$$
$$= \{z \mid z = g(t,x,p,v),\ (p,v) \in \Gamma \times V(t,x)\} \subset E_n ,$$
$$R(t,x) = \widetilde{g}[t,x,\Gamma \times V(t,x)]$$
$$= \{\widetilde{z} = (z^0, z) \mid \widetilde{z} = \widetilde{g}(t,x,p,v),\ (p,v) \in \Gamma \times V(t,x)\} \subset E_{n+1} .$$

Since

$$\widetilde{R}(t,x) = \{z = (z^0, z) \mid \widetilde{z} = \sum_{j=1}^{\gamma} p_j\ \widetilde{f}(t,x,u^{(j)}),$$
$$p \in \Gamma,\ u^{(j)} \in U(t,x),\ j = 1, \dots, \gamma\},$$

with $\gamma \geq n+1$, we see that $\widetilde{R}(t,x)$ is the convex hull of the set $\widetilde{Q}(t,x)$ in E_{n+1}, and hence $\widetilde{R}(t,x)$ is always convex. For weak solutions there is no reason to consider sets analogous to the sets $\widetilde{\widetilde{Q}}(t,x)$. Any usual admissible pair $[x(t),\ u(t)]$ can be thought of as a generalized element $[y(t), p(t),\ v(t)]$ by taking $p_j(t) = 1/\gamma$, $j = 1, \dots, \gamma$, $u^{(j)}(t) = u(t)$, and then $y = x$, $y^o = x^o$, $J = I = x^o(t_2)$.

Let Ω be the class of all admissible pair, say $[x = (x^0, x),\ u(t)]$, satisfying the usual differential equations, constraints and boundary conditions, and let Ω^* be the class of all generalized elements $[\widetilde{y}(t) = (y^0, y),\ p(t),\ v(t)]$ satisfying the corresponding differential equations and constraints, and the same boundary conditions. As mentioned above we have $\Omega \subset \Omega^*$. If

$$i = \inf_{\Omega} I[x, u], \qquad j = \inf_{\Omega^*} J[y, p, v],$$

then $\Omega \subset \Omega^*$ implies $i \geqq j$.

14. Property (P) .

It is a fairly general phenomenon that generalized trajectories and corresponding values of J can be approached by means of usual trajectories and corresponding values of I, so that $i = j$. We shall say that property (P) holds whenever $i = j$.

(i) (R. V. Gamkrelidze [3]). Under the hypotheses that $A = E_1 \times E_n$, that $U(t, x)$ depends on t only, that $U(t)$ is compact for every t, that $U(t)$ is an upper semicontinuous function of t, that $\tilde{f}$ satisfies a Lipschitz condition, and that Ω is the class of all admissible pairs $x(t)$, $u(t)$ (satisfying the differential equations the constraints, and the given boundary conditions), property (P) holds.

We omit the proof. In the more general situation considered in the present paper [A any closed set, $U(t, x)$ depending both on t and x, $U(t, x)$ closed, $U(t, x)$ satisfying condition (U), Ω any given class of admissible pairs $x(t)$, $u(t)$] a proof of property (P), that is, that $i = j$, is much more difficult. Particularly the case in which $U(t, x)$ depends on both t and x gives rise to a number of difficulties. Nevertheless, we were able to prove property (P) under a set of simple requirements which are satisfied in most cases and which are easier to verify than property (P). (see [1e]).

Property (P) has been proved also by J. Warga [19], T. Wazewski [20], and A. Turowicz [18] under assumptions different from those of R. V. Gamkrelidze [(i) above] and ours.

Property (P) is not valid in general as the following simple example shows. Take $m = 1$, $n = 2$, U made up of only two points $u = -1$ and $u = 1$, $A = E_1 \times A_o$, where E_1 is the t-axis, and A_o is made up of the three sides of the triangle of vertices $(0,0)$, $(1,1)$, $(2,0)$ of the xy-plane E_2, $I = \int_{t_1}^{t_2} y|u|dt$, differential system $dx/dt = 1$, $dy/dt = u$, and boundary conditions $t_1 = 0$, $x(0) = y(0) = 0$, $x(t_2) = 2$, $y(t_2) = 0$. Then Ω is made up of only one element: $[x(t), y(t), u(t),\ 0 \leq t \leq 2]$ with $x(t) = t$, $y(t) = t$, $u(t) = 1$ if $0 \leq t \leq 1$, $x(t) = t$, $y(t) = 2-t$, $u(t) = -1$ if $1 \leq t \leq 2$, and for this only element $I = 1$. On the other hand, Ω^* contains other elements, in particular, $[p_1(t) = p_2(t) = 1/2,\ u^{(1)}(t) = 1,\ u^{(2)}(t) = -1,\ x(t) = t,\ y(t) = 0, 0 \leq t \leq 2]$ for which $J = 0$. Thus $i = 1$ and $j = 0$. Another example has been given by A. Plis [9].

Analogously; if we denote by Ω_o the class of all usual admissible pairs $[x(t) = (x^o, x),\ u(t)]$ of Ω satisfying the given inequalities

$$\int_{t_1}^{t_2} \left| \frac{dx^i}{dt} \right|^p dt \leqq N_i, \quad i = 1, \ldots, n,$$

for certain constants $N_i > 0$, $p > 1$, we shall denote by Ω_o^* the class of all generalized elements $[\tilde{y}(t) = (y^o, y),\ p(t)]$ of Ω^* satisfying the same inequalities. Then $\Omega_0 \subset \Omega_0^*$. If i_0 and j_0 denote the the infimum of I in Ω_0 and the infimum of J in Ω_0^*, then again we have $i_0 \geqq j_0$. We shall say that property (P_0) holds whenever $i_0 = j_0$. We have proved that simple requirements analogous to the ones for (P) assure that also property (P_0) holds (see [1 e]).

For definitions of generalized or weak solutions slightly different from the one of R. V. Gamkrelidze and above, we refer to L. C. Young [21], E. J. McShane [6], J. Warga [19], and T. Wazewski [20].

15. Existence theorem

Existence Theorem VI. (Existence theorem for weak solutions) (L. Cesari [1e]). Let A be a compact subset of the tx-space $E_1 \times E_n$, let U(t, x) be a closed subset of E_m for every $(t, x) \in A$, and let $\tilde{f}(t, x, u) = (f_0, \ldots, f_n)$ be a continuous vector function on the set M of all (t, x, u) with $(t, x) \in A$, $u \in U(t, x)$. Let us assume that there is some continuous scalar function $\varphi(\zeta)$, $0 \leqq \zeta \leqq +\infty$, with $\varphi(\zeta)/\zeta \to +\infty$ as $\zeta \to +\infty$, such that $f_0(t, x, u) \geqq \varphi(|u|)$ for all $(t, x, u) \in M$, and that there are constants $C, D \geqq 0$ such that $|f(t, x, u)| \leqq C + D|u|$ for all $(t, x, u) \in M$. Let B be a closed subset of the $t_1 x_1 t_2 x_2$-space E_{2n+2}. Let us assume that U(t, x) satisfies property (U) in A, that $\tilde{R}(t, x)$ satisfies property (Q) in A, that property (P) holds, and that Ω is not empty. Then the infimum i of I [x, u] in Ω is attained by a weak solution (that is, i is attained by J [x, p, v] in the class Ω^*).

When A is not compact, but closed, then the theorem still holds under the additional hypotheses stated at the end of Existence Theorem 1.

Existence Theorem VII. (Existence theorem for weak solutions)

(L. Cesari [1e]). Let A be a compact subset of the tx-space $E_1 \times E_n$, and for every $(t,x) \in A$ let $U(t,x)$ be a closed subset of the u-space E_m. Let $\tilde{f}(t,x,u) = (f_0, f_1, \ldots, f_n)$ be a continuous vector function on the set M of all (t,x,u) with $(t,x) \in A$, $u \in U(t,x)$. Let B be a closed subset of the $t_1 x_1 t_2 x_2$-space E_{2n+2}. Let us assume that the set $U(t,x)$ satisfies property (U) in A, and that the set $\tilde{R}(t,x)$ satisfies property (Q) in A. Let us assume that $f_0(t,x,u) \geqq -G$ for some constant $G \geqq 0$ and all $(t,x,u) \in M$. Let Ω_0 be the class of all admissible pairs $[x(t), u(t)]$ of Ω satisfying the inequalities

$$\int_{t_1}^{t_2} \left| \frac{dx^i}{dt} \right|^p dt \leqq N_i, \quad i = 1, \ldots, n,$$

for some constants $N_i \geqq 0$, $p > 1$, and let Ω^* be the analogous subclass of all generalized elements $[\tilde{y}(t) = (y^0, y), p(t), v(t)]$ of Ω^* satisfying the same inequalities. Assume that Ω_0 is not empty, and property (P_0) holds. Then the infimum i of $I[x,u]$ in Ω_0 is attained by a weak solution (that is, i is attained by $J[y,p,v]$ in Ω_0^*).

When A is not compact, but A is closed, then Theorem VII still holds under the additional hypotheses stated at the end of Existence Theorem II. As for Theorem II, every inequality (29) which is a consequence of an inequality of the form $I \leqq N$ can be disregarded.

16. An example.

Let us consider the following simple Lagrange problem with m = 1,

$n = 2$, $A = E_2$, U made up of the points $u = 1$ and $u = -1$, differential equations $dx/dt = 1$, $dy/dt = u \in U$, and cost functional

$$(30) \qquad I[x, y, u] = \int_0^1 y^2 dt,$$

and boundary conditions $x(0) = y(0) = 0$, $x(1)=1$, $y(0) = 0$.

Let Ω denote the class of all pairs of AC functions $x(t)$, $y(t)$, $0 \leq t \leq 1$, satisfying the differential equations and constraints above and i be the infimum of $I[x, u]$ in Ω. Then $i \geq 0$. On the other hand, if we consider the sequence $x_k(t)$, $y_k(t)$, $0 \leq t \leq 1$, $k = 1, 2, \ldots$, defined by $x_k(t) = t$, $y_k(t) = t-ik^{-1}$, $u_k(t) = 1$ for $ik^{-1} \leq t \leq ik^{-1} + (2k)^{-1}$: $x_k(t) = t$, $y_k(t) = (i+1)k^{-1} - t$, $u_k(t) = -1$ for $ik^{-1} + (2k)^{-1} \leq (i+1)k^{-1}$, then $I[x_k, y_k, u_k] \to 0$ as $k \to \infty$, and hence $i \leq 0$. Thus $i = 0$. Now $I[x, y, u]$ cannot take the value zero in Ω since $I = 0$ implies $y = 0$, $u = 0$ a.e. in $[0, 1]$, a contradiction.

The generalized problem corresponding to (30) with $\gamma = 2$, taking $u^{(1)} = 1$, $u^{(2)} = -1$, $w = p_1 u^{(1)} + p_2 u^{(2)}$, is a Lagrange problem with $m = 1$, $n = 2$, $A = E_2$, $W = [w \mid -1 \leq w \leq 1]$, differential equations $dx/dt = 1$, $dy/dt = w \in W$, and same cost functional (30). Then $j = 0$, and $x(t) = t$, $y(t) = 0$, $w(t) = 0$ is obviously an optimal generalized solution corresponding to a random combination of the two strategies $u^{(1)}(t) = 1$ and $u^{(2)}(t) = -1$, $0 \leq t \leq 1$, with probabilities $p_1 = 1/2$, $p_2 = 1/2$. All conditions of Existence Theorem VI are here satisfied.

L. Cesari

Lecture 6. A system of partial differential equations in Sobolev spaces.

17. Notations.

Let G be a bounded open subset of the t-space E_ν, $t = (t^1, \dots, t^\nu)$, let $x = (x^1, \dots, x^n)$ denote a vector variable in E_n, and $u = (u^1, \dots, u^m)$ a vector variable in E_m. As usual we shall denote by $c\ell\, G$, and by bd $G = \partial G$ the closure and the boundary of G. For every $t \in c\ell\, G$, let A(t) be a given nonempty subset of E_n, and let A be the set of all (t, x) with $t \in c\ell G$, $x \in A(t)$. For every $(t, x) \in A$, let U(t, x) be a subset of E_m, and let M be the set of all (t, x, u) with $(t, x) \in A$, $u \in U(t, x)$.

The set A defined above is a subset of $E_\nu \times E_n$, and its projection on E is $c\ell\, G$. The set M defined above is a subset of $E_\nu \times E_n \times E_m$ and its projection on $E_\nu \times E_n$ is A.

We shall assume below that G is bounded by a surface S which is a regular boundary in the sense of S. L. Sobolev [15, Chap. 1, § 10, p. 72, Remark], and for the sake of simplicity we shall say that G is of class K_ℓ. Thus, S can be decomposed into a finite number of manifolds $S_1, \dots, S_J$ of dimension n-1 (and corresponding boundaries), each S_j, j = 1, ..., J, having the property that it can be transformed into a hyperplane π_j by means of a transformation of coordinates T_j defined on a part G_j of G and continuous with continuous derivatives up to ℓ^{th} order, j = 1, ..., J.

We shall denote by $x(t) = (x^1, \dots, x^n)$, $u(t) = (u^1, \dots, u^m)$, $t \in G$ vector functions of t in G. For every i = 1, ..., n, we shall denote by $\{\alpha\}$ a given finite system of indices $\alpha = (\alpha_1, \dots, \alpha_\nu)$, $\alpha_j \geq 0$ integers, $0 \leq |\alpha| \leq \ell_1$, with $|\alpha| = \alpha_1 + \dots + \alpha_\nu$, $\ell_i \leq \ell$. We shall assume that each component $x^i(t)$ of $x(t)$ is L_{p_i}-integrable in G and possesses the generalized partial derivatives $D^\alpha x^i(t)$ of the orders $\alpha \in \{\alpha\}_i$, all L_{p_i}-integrable in G for certain $p_i \geq 1$, $i = 1, \dots, n$. We shall assume that each component $u^j(t)$ of $u(t)$ is measurable in G.

Let N denote the total number of indices α contained in the n systems $\{\alpha\}_i$ $i = 1, \dots, n$, and let $f(t, x, u) = (f_{i\alpha})$ denote an N-vector function, whose components are real valued functions $f_{i\alpha}(t, x, u)$ defined on M. We shall consider the system of N partial differential equations in G:

$$D^\alpha x^i = f_{i\alpha}(t, x, u), \quad \alpha \in \{\alpha\}_i \quad i = 1, \dots, n,$$

or briefly

$$D^\alpha x = f(t, x, u).$$

We are interested in pairs $[x(t), u(t)]$ of vector functions $x(t)$, $u(t)$, $t \in G$, as above, satisfying the constraints

$$(t, x(t)) \in A, \quad u(t) \in U(t, x(t)), \quad \text{a.e. in } G,$$

and satisfying the system of partial differential equations

$$D^\alpha x^i(t) = f_{i\alpha}(t, x(t), u(t)), \quad \text{a.e. in } G, \alpha \in \{\alpha\}_i = 1, \dots, n,$$

or briefly

$$D^\alpha x(t) = f(t, x(t), u(t)), \quad \text{a.e. in } G.$$

Lecture 7. A closure theorem in Sobolev spaces.

18. Closure Theorem III (in Sobolev spaces).

Let G be a bounded open set of the t-space E_ν of some class K_ℓ, $\ell \geq 1$, let $A(t)$ be a nonempty subset of the x-space E_n defined for every $t \in cl\, G$, and assume that the set A of all $(t, x) \in E_\nu \times E_n$ with $t \in cl\, G$, $x \in A(t)$, be compact. Let $U(t, x)$ be a nonempty compact subset of the u-space E_m defined for every $(t, x) \in A$, and assume that $U(t, x)$ is an upper semicontinuous function of (t, x) in A. Let M be the set of all $(t, x, u) \subset E_\nu \times E_n \times E_m$ with $(t, x) \in A$, $u \in U(t, x)$. Let $\{\alpha\}_i$ be a finite system of indices $\alpha = (\alpha_1, \dots, \alpha_\nu)$, $1 \leq |\alpha| \leq \ell_i$, with $\ell_i \leq \ell$, defined for every $i = 1, \dots, n$, and let N be the total number of elements $\alpha \in \{\alpha\}_i$, $i = 1, \dots, n$. Let $f(t, x, u) = (f_{i\alpha})$ $(\alpha \in \{\alpha\}_i$, $i = 1, \dots, n)$, be a continuous N-vector function on the set M, and assume that the set $Q(t, x) = f(t, x, U(t, x))$ be a convex subset of E_N for every $(t, x) \in A$. Let $x_k(t)$, $u_k(t)$, $t \in G$, $k = 1, 2, \dots$, be a sequence of pairs of vector functions $x_k(t) = (x_k^1, \dots, x_k^n)$, $u_k(t) = (u_k^1, \dots, u_k^m)$, $x_k^i(t) \in L_1(G)$, $u^i(t)$ measurable in G, and let us assume that each component $x_k^i(t)$ of $x_k(t)$ possesses the generalized derivatives $D^\alpha x_k^i(t)$, $t \in G$, for $\alpha \in \{\alpha\}_i$, $D^\alpha x_k^i(t) \in L_1(G)$, such that

$$(30) \qquad (t, x_k(t)) \in A, \quad u_k(t) \in U(t, x_k(t)), \quad \text{a.e. in } G,$$

$$(31) \qquad D^\alpha x_k^i(t) = f_{i\alpha}(t, x_k(t), u_k(t)) \quad \text{a.e. in } G,$$

for all $\alpha \in \{\alpha\}_i$, $i = 1, \dots, n$, $k = 1, 2, \dots$. Let $x(t) = (x^1, \dots, x^n)$, $t \in G$, be a vector function whose components $x^i(t) \in L_1(G)$ possess generalized partial derivatives $D^\alpha x^i(t), \alpha \in \{\alpha\}_i$, $i = 1, \dots, n$, and assume that

(32) $$x_k^i(t) \longrightarrow x^i(t) \text{ in } L_1(G) \text{ as } k \longrightarrow \infty,$$

(33) $$\int_R D^\alpha x_k^i(t)dt \longrightarrow \int_R D^\alpha x^i(t)\,dt \quad \text{as} \quad k \longrightarrow \infty$$

for every $\alpha \in \{\alpha\}_i$, $i = 1, \dots, n$, and every interval $R \subset G$.

Then there is a measurable vector function $u(t) = (u^1, \dots, u^m)$, $t \in G$, such that

(34) $$(t, x(t)) \in A, \quad u(t) \in U(t, x(t)) \quad \text{a.e. in } G,$$

(35) $$D^\alpha x^i(t) = f_{i\alpha}(t, x(t), u(t)) \quad \text{a.e. in } G,$$

for all $\alpha \in \{\alpha\}_i$, $i = 1, \dots, n$.

If the functions $x_k^i(t)$, $k = 1, 2, \dots$, and $x^i(t)$ belong to a Sobolev's class $W_{p_i}^{\ell_i}(G)$ for some $p_i \geq 1$, and $x_k^i(t) \longrightarrow x^i(t)$ weakly in $W_{p_i}^{\ell_i}(G)$ then (32) and (33) are certainly satisfied. Indeed all derivatives $D^\alpha x_k^i(t)$ of orders $|\alpha| = \ell_i$ converge to $D^\alpha x^i(t)$ weakly in $L_{p_i}(G)$, and the functions $x_k^i(t)$ and all derivatives $D^\alpha x_k^i(t)$ of orders $0 \leq |\alpha| \leq \ell_i - 1$ converges to $D^\alpha x_k^i$ strongly in $L_{p_i}(G)$, and thus (32) and (33) certainly hold [7].

<u>Proof.</u> By hypothesis, A is a compact subset of $E_\nu \times E_n$. Hence, its projection on the x-space E_n is compact, and therefore bounded, say $|t| \leq M_1$, $|x| \leq M_1$ for every $(t, x) \in A$. Also, $U(t, x)$ is a compact subset of E_m for every (t, x) of the compact

set A , and an upper semicontinuous function of (t, x) in A. Therefore, M is a compact subset of $E_\nu \times E_n \times E_m$ (Cfr. [1e]). Since f is continuous on the compact set M, then f is bounded on M, say $|f| \leq M_o$, and $|f_{i\alpha}| \leq M_{i\alpha}$ where M_o, $M_{i\alpha}$ are constants $\alpha \in \{\alpha\}_i$, $i = 1, \ldots, n$. Also, the convex set $Q(t, x) = f(t, x, U(t, x))$ is compact for every $(t, x) \in A$, and an upper semicontinuous function of (t, x) in A (Cfr. [1 e]).

For every $t_0 \in G$, $t_0 = (t_{01}, \ldots, t_{0\nu})$, let us denote by δ_0 the distance of t_0 from ∂G, and by $q = q_h = [t_0, t_0 + h]$ the closed hypercube $q = [t_{0j} \leq t \leq t_{0j} + h,\ j = 1, \ldots, \nu]$ where h denotes a positive number. For simplicity we denote by h also the ν-vector $(h_1, \ldots, h_\nu)$. For $0 \leq h \leq \delta_0/\nu$, we have $q_h \subset G$.

For almost every $t_0 \in G$ we have

$$(36) \qquad x^i(t_0) = \lim_{h \to 0} h^{-\nu} \int_q x^i(t)dt ,$$

$$(37) \qquad D^\alpha x^i(t_0) = \lim_{h \to 0} h^{-\nu} \int_q D^\alpha x^i(t)dt, \quad \alpha \in \{\alpha\}_i, \ i = 1, \ldots, n.$$

Also, for almost $t \in G$ we have $(t, x_k(t)) \in A$ for all $k = 1, 2, \ldots$. Finally, the convergence $x^i_k(t) \to x^i(t)$ in $L_1(G)$ as $k \to \infty$ implies convergence in measure in G, and hence there is a subsequence $[x_{k_r}(t)]$ which converges pointwise almost everywhere in G. Let G_0 be the subset of all $t \in G$, where (36) and (37) hold, where $(t, x_k(t)) \in A$ for all k, and $x_{k_r}(t) \longrightarrow x(t)$ as $r \longrightarrow \infty$. Then G_0 is measurable and meas G_0 = meas G.

Let us prove that $(t, x(t)) \in A$ a.e. in G. Indeed, $(t, x_{k_r}(t)) \in A$

for all $t \in G_0$ and r, $x_{k_r}(t) \longrightarrow x(t)$ as $r \longrightarrow \infty$, and A is closed. Then $(t, x(t)) \in A$ for every $t \in G_0$, that is, a.e. in G.

Let t_0 be any point of G_0, hence $(t_0, x(t_0)) \in A$. Then, the sets $U(t_0, x_0)$, $Q(t_0, x_0) = f(t_0, x_0, U(t_0, x_0))$, where $x_0 = x(t_0)$, are defined. Since $U(t, x)$ and $Q(t, x)$ are upper semicontinuous functions of $(t, x) \in A$, given $\epsilon > 0$, there is some $\delta = \delta(t_0, x_0, \epsilon) > 0$ such that $|t - t_0| < \delta$, $|x - x_0| \leq \delta$, $(t, x) \in A$, implies $U(t, x) \subset [U(t_0, x_0)]_\epsilon$ $Q(t, x) \subset [Q(t_0, x_0)]_\epsilon$.

Because of the convergence $x_k(t) \longrightarrow x(t)$ in $L_1(G)$, and hence in measure, and consequent pointwise convergence $x_{k_r} \longrightarrow x(t)$ every- where in G_0 with meas G_0 = meas G, we know that there are closed sets C_λ, $\lambda = 1, 2, \ldots$, with $C_\lambda \subset G_0$, $C_\lambda \subset C_{\lambda+1}$, meas $C_\lambda >$ meas $G_0 - 1/\lambda$, such that $x_{k_r}(t)$, $x(t)$ are continuous on C_λ, and $x_{k_r}(t) \rightrightarrows x(t)$ uniformly on C_λ as $r \longrightarrow \infty$, for every $\lambda = 1, 2, \ldots$. Since G is bounded and $C_\lambda \subset G_0 \subset G$, each set C_λ is compact, and $x(t)$, $x_{k_r}(t)$, $r = 1, 2, \ldots$, are continuous, uniformly continuous, and equicontinuous on each C_λ.

Let λ be any fixed integer with $\lambda > (\text{meas } G)^{-1}$, hence meas $C_\lambda > 0$. Let $\eta = \delta/2$. Then, there is some $\delta' = \delta'(\delta, \lambda) > 0$ such that $|t - t'| \leq \delta'$, $t, t' \in C_\lambda$ implies $|x(t) - x(t')|$, $|x_{k_r}(t) - x_{k_r}(t')| \leq \eta$ for every $r = 1, 2, \ldots$. Also, there exists some $k = k(\delta, \lambda)$ such that $|x_{k_r}(t) - x(t)| \leq \eta$ for all $t \in C_\lambda$, $k_r \geq k(\delta, \lambda)$. For almost every point $t_0 \in C_\lambda$ we have

$$(38) \qquad \lim_{h \to 0} h^{-\nu} \text{ meas}(q \cap C_\lambda) = 1 .$$

Let C'_λ bet he subset of C_λ where this occurs. Then C'_λ is measurable, $C'_\lambda \subset C_\lambda \subset G_0 \subset G$, meas C'_λ = meas C_λ > meas $G - 1/\lambda$ >0.

Let t_0 be any point of C'_λ. Let us fix $h > 0$ so small that $h \leq \varepsilon$. $h \leq \delta/\nu$, $h < \delta'/\nu$, and

$$\left| D^\alpha x^i(t_0) - h^{-\nu} \int_q D^\alpha x^i(t)\, dt \right| < \varepsilon/N, \alpha \in \{\alpha\}_i \quad , \; i=1,\dots,n,$$

(39)

$$h^{-\nu} \text{ meas } (q \cap C_\lambda) > 1 - (\varepsilon/M_0).$$

This is possible because of (37) and (38).

Having so fixed $h > 0$, (33) holds for $R = q = q_h$. Then we can determine an integer $k = k_r$ so large that $k \geq k(\delta, \lambda)$ and

$$(40) \quad \left| h^{-\nu} \int_q D^\alpha x^i(t)dt - h^{-\nu} \int_q D^\alpha x^i_k(t)dt \right| < \varepsilon/N, \alpha \in \{\alpha\}_i, \; i=1,\dots,n.$$

In vector form (39), (40), and (31) imply

$$\left| D^\alpha x(t_0) - h^{-\nu} \int_q f(t, x_k(t), u_k(t))dt \right| \leq$$

$$\leq \left| D^\alpha x(t_0) - h^{-\nu} \int_q D^\alpha x(t)dt \right| + \left| h^{-\nu} \int_q D^\alpha x(t)dt - h^{-\nu} \int_q f(t, x_k(t), u_k(t))dt \right|$$

$$(41) \qquad \leq N(\varepsilon/N) + N(\varepsilon/N) = 2\varepsilon.$$

Let H be the set $H = q \cap C_\lambda$. Then $|H|$ = meas $H \geq (1 - \varepsilon/M_0)h^\nu$, $1 \geq h^{-\nu} |H| \geq 1 - (\varepsilon/M_0)$, meas $(q - H) < h^\nu \varepsilon/M_0$.

For $t \in H$, we have

$$|t - t_0| \leq \nu h \leq \delta,$$

$$|x_k(t) - x(t_0)| \leq |x_k(t) - x_k(t_0)| + |x_k(t_0) - x(t_0)| \leq \eta + \eta = \delta,$$

and hence

$$(42) \qquad f(t, x_k(t), u_k(t)) \in Q(t, x_k(t)) \subset [Q(t_0, x_0)]_\varepsilon, \qquad t \in H$$

Then, for the average of these values in H, we have also

$$|H|^{-1} \int_H f(t, x_k(t), u_k(t))dt \in [Q(t_0, x_0)]_\epsilon ,$$

since the last set is convex and closed. On the other hand, since $|f(t, x_k(t), u_k(t))| \leq M_0$ for every t, we have

$$\left| |H|^{-1} \int_H f(t, x_k(t), u_k(t))dt - h^{-\nu} \int_H f(t, x_k(t), u_k(t))dt \right|$$

$$= \left| (1 - h^{-\nu}|H|) \ |H|^{-1} \int_H f(t, x_k(t), u_k(t))dt \right|$$

$$\leq (\epsilon / M_0) M_0 = \epsilon ,$$

and hence, by comparison with (42) we have

$$(43) \qquad h^{-\nu} \int_H f(t, x_k(t), u_k(t))dt \in [Q(t_0, x_0)]_{2\epsilon} .$$

On the other hand

$$\left| h^{-\nu} \int_q f(t, x_k(t), u_k(t))dt - h^{-\nu} \int_H f(t, x_k(t), u_k(t))dt \right| =$$

$$= \left| h^{-\nu} \int_{q-H} f(t, x_k(t), u_k(t))dt \right| \leq h^{-\nu} M_0 \text{meas } (q-H)$$

$$\leq h^{-\nu} M_0 (h^{\nu} \epsilon / M_0) = \epsilon ,$$

and finally, by comparison with (43), we have

(44) $$h^{-\nu}\int_q^{r} f(t, x_k(t), u_k(t))\, dt \in [Q(t_0, x_0)]_{3\epsilon}.$$

By (41) we conclude that

$$D^\alpha x(t_0) \in [Q(t_0, x_0)]_{5\epsilon} = [Q(t_0, x(t_0)]_{5\epsilon}.$$

This relation is valid for every $\epsilon > 0$,hence

$$D^\alpha x(t_0) = \bigcap_\epsilon [Q(t_0, x(t_0))]_{5\epsilon} = Q(t_0, x(t_0)) ,$$

since the last set is compact . This proves that there is some $\bar{u}(t_0) \in U(t_0, x(t_0))$ such that

$$D^\alpha x(t_0) = f(t, x(t_0), \bar{u}(t_0)) .$$

This holds for every $t_0 \in C'_\lambda$. Since meas C'_λ >meas $G - 1/\lambda$, and λ can be taken as large as we want , we conclude that there is at least one function $\bar{u}(t)$ defined almost everywhere in G satisfying (34) and (35) . To prove that there is at least one such function u(t) which is also measurable in G we can repeat the usual Filippov's argument for the case $\nu = 1$. The reasoning is the same (Cfr. [1e]).

Remark . We mention a variant of closure theorem I which is of some interest. We may assume that G is made up of components $G_1, \dots, G_\mu$, and in each of these there is a different system of $\{\alpha\}_i$, $i = 1, \dots, n$, and of functions $f_{i\alpha}$. Another situation is of interest. Assume that G_s, $s = 1, \dots, \mu$, are finitely many open bounded subsets of E_ν , and that for each G_s there is a given set $\{\alpha\}_{is}$,

$i = 1, \dots, n$, $s = 1, \dots, \mu$, and a system $f_{i\alpha}$ of functions f. Now let us consider all possible nonempty intersections $G_r = G_{s_1} \cap G_{s_2} \cap \dots \cap G_{s_p}$ of p of the sets G_s, $1 \leq p \leq \mu$. These sets G_r, $r = 1, \dots, N$, are finitely many and each is a nonempty bounded open subset of E_ν. For each of these N sets $G_r = G_{s_1} \cap \dots \cap G_{s_p}$ we shall consider the vector function $f^{(r)}(t, x, u)$ whose components are all those of the functions $f = (f_{i\alpha})$, $\alpha \in \{\alpha\}_{is}$, $i = 1, \dots, n$, $s = s_1, \dots, s_p$. We shall then require that, for each r, the set $Q^{(r)}(t, x) = f^{(r)}(t, x, U(t, x))$ is a convex subset of the relative Euclidean space E_{M_r}. Here M_r is the total number of all distinct indices $\alpha \in \{\alpha\}_{is}$, $i = 1, \dots, n$, $s = s_1, \dots, s_p$. In other words, in each G_s we have a different system of partial differential equations $D^\alpha x^i = f_{i\alpha}$, and in each nonempty intersection $G_r = G_{s_1} \cap \dots \cap G_{s_p}$ we consider the logical union of the various differential systems, and the hypothesis "$Q^{(r)}(t, x)$ convex" corresponds to the usual hypothesis for these composite differential systems. A further extension is obtained when we assume that the functions $f_{i\alpha}$ are sectionally continuous in each set G_r, but, for instance, coincide on each set G_r with functions which are continuous on the closure of G_r. These variants of Closure Theorem I are proved exactly by the same argument.

Lecture 8. Existence theorems for Pontryagin's problems in Sobolev spaces.

19. More notations for the existence theorems.

We shall use the same notations as in No. 17. Beside the N-vector $f(t, x, u) = (f_{i\alpha})$, we now consider a scalar function $f_0(t, x, u)$ continuous on M, and we denote by $\tilde{f}(t, x, u)$ the (N + 1) - vector function $\tilde{f}(t, x, u) = (f_0, f_{i\alpha})$ continuous on M. Concerning the n-vector $x = (x^1, \dots, x^n)$ we shall require that each variable x^i belongs to a class $W_{p_i}^{\ell_i}(G)$ for given $\ell_i \geq 1$, $p_i > 1$, $i = 1, \dots, n$. By force of Sobolev's imbedding theorems each variable x^i and each of its derivative $D^\alpha x^i$ of order $0 \leq |\alpha| \leq \ell_i - 1$, has boundary values $\phi_\alpha^i(t)$ defined almost everywhere on S, each of class L_{p_i} on S.

We shall now require a set (B) of boundary conditions involving the boundary values of the functions x^i and of their derivatives $D^\alpha x^i$, $0 \leq |\alpha| \leq \ell_i - 1$.

On the boundary conditions (B) we assume the following closure property:

(P_1) If $x(t) = (x^1, \dots, x^n)$, $x_k(t) = (x_k^1, \dots, x_k^n)$, $k = 1, 2, \dots$, are vector functions whose components x^i, x_k^i belong to the Sobolev's class $W_p^{\ell_i}(G)$, if $D^\beta x_k^i(t) \longrightarrow D^\beta x^i(t)$ as $k \to \infty$, strongly in $L_P(G)$ for every β with $0 \leq |\beta| \leq \ell_i - 1$, if $D^\beta x_k^i(t) \to D^\beta x^i(t)$ as $k \to \infty$ weakly in $L_p(G)$ for every β with $|\beta| = \ell_i$, and if the boundary values $\phi_{k\alpha}^i(t)$ of $x_k^i(t)$ on ∂G satisfy boundary condition (B), then the boundary values $\phi_\alpha^i(t)$ of $x^i(t)$ on ∂G satisfy condition (B).

For instance, if the boundary condition (B) are defined by stating that some of the boundary values $\phi_{k\alpha}^{(i)}(t)$ coincide with preassigned continuous functions $\psi_\alpha^{(i)}(t)$, $t \in G$, then by force of Sobolev's imbedding theorems [15] we know that property (P_1) is valid.

We shall need later on a further property of boundary conditions (B):

(P_2) If $x(t) = (x^1, \ldots, x^n)$, $t \in G$, denotes any vector function satisfying boundary conditions (B), whose components $x^i(t) \in W_{p_i}^{\ell_i}(G)$, $p_i > 1$, $\ell_i \geq 1$,

$$\int_G \left| D^\beta x^i(t) \right|^{p_i} dt \leq M_\beta$$

for all $\beta = (\beta_1, \ldots, \beta_\nu)$. $|\beta| = \ell_1$, $i = 1, \ldots, n$, and constants M_β, then there are constants M_α such that

$$\int_G \left| D^\alpha x^i(t) \right|^{p_i} dt \leq M_\alpha$$

for all $\alpha = (\alpha_1, \ldots, \alpha_\nu)$, $0 \leq |\alpha| \leq \ell_i - 1$, $i = 1, \ldots, n$, where the constants M_α depend only on p_i, ν, all M_β, G, and boundary conditions (B), but not on the vector function $x(t)$ as above.

For instance, the boundary conditions (B) defined by preassigning the continuous boundary value functions $\phi^\alpha(t)$, $t \in \partial G$, of all derivatives $D^\alpha x^i(t)$, $\alpha = (\alpha_1, \ldots, \alpha_\nu)$, $0 \leq |\alpha| \leq \ell_i^{-1}$ $i = 1, \ldots, \nu$, satisfy condition (P_2).

20. An existence theorem for multidimensional problems of optimal control.

Existence Theorem VIII. Let G be a bounded open set of class K_ℓ of the t-space E_ν, let $A(t)$ be a nonempty subset of the x-space E_n defined for every $t \in$ cl G, and assume that the set A of all $(t, x) \in E_\nu \times E_n$ with $t \in$ cl G, $x \in A(t)$, be compact. Let

$U(t, x)$ be a nonempty compact subset of the u-space E_m defined for every $(t, x) \in A$, and assume that $U(t, x)$ is an upper semicontinuous function of (t, x) in A. Let M be the set of all (t, x, u) $\in E_\nu \times E_n \times E_m$ with $(t, x) \in A$, $u \in U(t, x)$. For every $i = 1, \dots, n$, let $\{\alpha\}_i$ be a given finite system of indices $\alpha = (\alpha_1, \dots, \alpha_\nu)$, $0 \leq |\alpha| \leq \ell_i \leq \ell$, and let N be the total number of elements $\alpha \in \{\alpha\}_i$, $i = 1, \dots, n$. Let $\widetilde{f}(t, x, u) = (f_o, f_{i\alpha})$, $(\alpha \in \{\alpha\}_i$, $i=1, \dots, n)$, be a continuous $(N + 1)$ - vector function on the set M, and assume that the set $\widetilde{\widetilde{Q}}(t, x) = [\widetilde{z} = (z^o, z) \mid z^o \geq f_o(t, x, u), z = f(t, x, u), u \in U(t, x)]$ is a convex subset of E_{N+1} for every $(t, x) \in A$. Let (B) be a system of boundary conditions satisfying properties (P_1) and (P_2). Let Ω be a nonempty complete class of pairs $x(t) = (x^1, \dots, x^n)$, $u(t) = (u^1, \dots, u^m)$, $t \in G$, $x^i(t) \in W^{\ell_i}_{p_i}(G)$ $p_i > 1$ $1 \leq \ell_i \leq \ell$, $i = 1, \dots, n$, $u^j(t)$ measurable in G, $j = 1, \dots, m$, satisfying (a) the constraints

$$(t, x(t)) \in A, \quad u(t) \in U(t, x(t)) \quad , \text{ a.e. in } G \ ;$$

(b) the system of partial differential equations

$$\bar{D}^\alpha x^i(t) = f_{i\alpha}(t, x(t), u(t)) \ , \text{a.e. in } G, \alpha \in \{\alpha\}_i, \ i = 1, \dots, n;$$

(c) boundary conditions (B) on the boundary ∂G, concerning the boundary values of each $x^i(t)$ and of its generalized partial derivatives $\bar{D}^\beta x^i(t)$ of orders $0 \leq |\beta| \leq \ell_i - 1$; (d) the finite system inequalities

(45) $$\int_G \left|\bar{D}^{\beta} x^i(t)\right|^{p_i} dt \leq M_{i\beta}\ ,\quad \beta = \{\beta\}_i\ i = 1, \ldots, n,$$

where $M_{i\beta}$ are given numbers, and $\{\beta\}_i$ contains at least all indices $\beta = (\beta_1, \ldots, \beta_\nu)$ with $|\beta| = \ell_i$ which are not already in $\{\alpha\}_i$, $i = 1, \ldots, n$. Then the cost functional

(46) $$I[x, u] = \int_G f_0(t, x(t), u(t))dt$$

possesses an absolute minimum in Ω.

If it happens that for any $M > 0$ sufficiently large and given i and $\beta \in \{\beta\}_i$ there is some $N_{i\beta}$ such that $I[x, u] \leq M$ implies $\int_G \left| D^{\beta} x^i(t)\right|^{p_i} dt \leq N_{i\beta}$, then the absolute minimum still exists even if we disregard the corresponding relation (45) in defining Ω (but the absolute minimum may change).

Proof of Existence Theorem VIII. By standard continuity argument the set M is compact. Then the continuous functions $f_0, f_1, \ldots, f_n$ are bounded on M, say $|f_0(t, x, u)| \leq M_0$, $|f_{i\alpha}(t, x, u)| \leq M_{i\alpha}$.

Thus, for every pair $x(t)$, $u(t)$ in Ω, $f_0(t, x(t), u(t))$ is measurable in G, $|f_0(t, x(t), u(t))| \leq M_0$, and the cost functional (46) exists as an L-integral and has a finite value.

Let $\tilde{U}(t, x)$ denote the set of all $\tilde{u} = (u^0, u) \in E_{m+1}$ with $M_0 \geq u^0 \geq f_0(t, x, u)$ and by $\tilde{\tilde{Q}}(t, x)$ the set of all $z = (z^0, z^1, \ldots, z^N) = (z^0, z) \in E_{N+1}$ with $M_0 \geq z^0 \geq f_0(t, x, u)$, $z = f(t, x, u)$, $u \in U(t, x)$, $(t, x) \in A$. If $\tilde{\tilde{f}}(t, x, \tilde{u}) = (\tilde{f}_0, f_1, \ldots, f_n) = (\tilde{f}_0, f)$ with $\tilde{f}_0 = u^0$, then obviously $\tilde{\tilde{Q}}(t, x) = \tilde{\tilde{f}}(t, x, \tilde{U}(t, x))$. We shall consider the auxiliary problem

$$J[x; \tilde{u}] = \int_G \tilde{f}_0(t, x, \tilde{u})dt = \int_G u^0 \, dt .$$

If $x(t)$, $u(t)$, $t \in G$, is an admissible pair for $I[x, u]$, and we take $u^0(t) = f_0(t, x(t), u(t))$ and $\tilde{u}(t) = (u^0, u)$, then $x(t)$, $\tilde{u}(t)$ is now an admissible pair for the auxiliary integral $J[x, \tilde{u}]$ and $J[x, \tilde{u}] = I[x, u]$. The converse is not true since $u^0(t) \geq f_0(t, x(t), u(t))$ and the sign $>$ may hold in a set of positive measure. On the other hand we have always

$$J[x, \tilde{u}] = \int_G \tilde{f}_0(t, x, \tilde{u})dt = \int_G u^0 dt \geq \int_G f_0(t, x, u)dt = I[x, u].$$

We have already denoted by Ω the class of all admissible pairs $x(t)$, $u(t)$ for the integral $I[x, u]$. We shall denote by $\tilde{\Omega}$ the class of all admissible pairs $x(t)$, $\tilde{u}(t)$ for the integral $J[x. \tilde{u}]$. If $x(t)$, $\tilde{u}(t)$ is in the class $\tilde{\Omega}$ then x, u is in the class Ω, and, as proved above, $J[x, \tilde{u}] \geq I[x, u]$. Also, if the pair x, $\tilde{u}$ is in $\tilde{\Omega}$, hence $\tilde{u} = (u^o, u)$, $u^o(t) \geq f_o(t, x(t), u(t))$, and we take $\bar{u}^o(t) = f_o(t, x(t), u(t))$, then the pair x, $\bar{u}$ is in $\tilde{\Omega}$, x, u is in Ω, and $J[x, \bar{u}] = I[x, u]$. This proves that $J[x, \tilde{u}]$ in $\tilde{\Omega}$, and $I[x, u]$ in Ω have the same infimum, and have or not absolute minimum together.

By standard continuity argument we can prove that $\tilde{U}(t, x)$ and $\tilde{\tilde{Q}}(t, x)$ are compact sets for every $(t, x) \in A$, and that $\tilde{U}(t, x)$ and $\tilde{\tilde{Q}}(t, x)$ are upper semicontinuous functions of (t, x) in A.

Let R be an interval $R = [a, b]$ containing $cl\,G$ in its interior, say $cl\,G \subset R^o \subset E_\nu$. It is not restrictive to assume $R = [0, b]$, where 0 and b represent they ν-vectors $(0, \dots, 0)$ and $(b, \dots, b)$. Let x^0 denote a new variable and $\tilde{x} = (x^0, x) = (x^0, x^1, \dots, x^n)$. Let $\ell_o = \nu$, and let $\bar{f}_o(t, x, u)$ be a new scalar function defined by

$$\bar{f}_o(t, x, u) = 0 \quad \text{for } t \in R - cl\,G,\ x \in E^n,\ u \in E^m,$$

$$\bar{f}_o(t, x, u) = f_o(t, x, u) \text{ for } t \in cl\,G,\ (t, x, u) \in M.$$

We shall denote by $x^0(t)$, $t \in R$, a new scalar function, which is L-integrable in R, possesses generalized partial derivative $\partial^\nu x^0 / \partial t^1 \dots \partial t^\nu$, or $D^\alpha x^0$ with $\alpha = (1, \dots, 1)$, which is also L_1-integrable in R, satisfies the partial differential equation

$$D^\alpha x^0 = \bar{f}_o(t, x(t), u(t)), \qquad \text{a.e. in } R,\ \alpha = (1, \dots, 1),$$

and the boundary conditions

$$x^0(t_1', 0) = 0,\ (t_1', 0) \in \partial R,\ i = 1, \dots, \nu,$$

where t_1' denotes the $(\nu - 1)$-vector $t_1' = (t_1, \dots, t_{i-1}, t_{i+1}, \dots, t_\nu)$. Then we have

$$x^0(b) = \int_R \bar{f}_0(t, x(t), u(t))dt = \int_G f_0(t, x(t), u(t))dt = I[x, u].$$

Here R is divided into the two parts $R - cl\,G$ and G, by the surface $S = \partial G$ of G.

For every pair $x(t)$, $u(t)$, $t \in G$, of the class Ω we have $I[x, u] = \int_G f_0 \geq -M_o$ meas G. Thus, if $i = \inf I[x, u]$ where Infimum is taken in Ω, whe conclude that i is finite. Let $x_k(t)$, $u_k(t)$, $t \in G$, $k = 1, 2, \dots$, be a sequence of pairs all in Ω with $I[x_k, u_k] \to i$ as $k \to \infty$. We can assume $i \leq I[x_k, u_k] \leq i + 1/k \leq i + 1$, $k = 1, 2, \dots$.

For every pair $x(t)$, $u(t)$, $t \in G$, of the class Ω, we have

$x^i(t) \in W_{p_i}^{\ell_i}(G)$, $i = 1, \dots, n$, and we may divide the derivatives $D^\alpha x^i$ of order $|\alpha| = \ell_i$ into two classes, one for which $\alpha \in \{\alpha\}_i$, and then $D^\alpha x^i(t) = f_{i\alpha}(t, x(t), u(t))$ a.e. in G, and hence $|D^\alpha x^i(t)| \leq M_{i\alpha}$, and then

$$\int_G |D^\alpha x^i(t)|^p \, dt \leq M_{i\alpha}^p \text{ meas } G = M'_{i\alpha} \quad ,$$

and the complementary class for which bounds are given in the form

$$\int_G |D^\alpha x^i(t)|^p dt \leq M'_{i\alpha}$$

Thus, for every pair $x(t)$, $u(t)$, $t \in G$, of the class Ω we have

$$\int_G |D^\alpha x^i(t)|^p \, dt \leq M'_{i\alpha}, |\alpha| = \ell_i, \; i=1, \dots, n.$$

By property (P_2) we conclude that there are constants M'_i such that

$$\|x^i\| = \left(\sum_{\rho=0}^{\ell_i} \sum_{|\alpha| = \rho} \int_G |D^\alpha x^i(t)|^p dt \right)^{1/p} \leq M'_1, \; i = 1, \dots, n.$$

Since any closed sphere in Sobolev spaces $W_p^{\ell_i}(G)$ is weakly compact, then by n successive extractions we can select a subsequence $[x_k(t)]$ for which each component $[x^i_{k_r}(t)]$ converges weakly toward a function $x^i(t)$, $t \in G$, also of class $W_p^{\ell_i}(G)$. Then, by property (P_1), the N-vector $x(t)$ satisfies boundary conditions (B) on ∂G, and by known properties of Sobolev's spaces we have

$$D^\alpha x^i_{k_r} \longrightarrow D^\alpha x^i(t) \text{ strongly in } L_p(G) \text{ as } r \longrightarrow \infty \text{ for } 0 \le |\alpha| \le \ell_i - 1,$$

$$D^\alpha x^i_{k_r} \longrightarrow D^\alpha x^i(t) \text{ weakly in } L_p(G) \text{ as } r \longrightarrow \infty \text{ for } |\alpha| = \ell_i,$$

where $i = 1, \dots, n$. The second line implies that for every interval $I \subset G$ we have also

$$\int_I D^\alpha x^i_{k_r}(t)\, dt \longrightarrow \int_I D^\alpha x^i(t) dt \quad \text{as } r \longrightarrow \infty \text{ for } |\alpha| = \ell_i,\ I \subset G.$$

Concerning the auxiliary variable x^o, we shall consider, for any pair $x(t)$, $u(t)$, $t \in G$, of the class Ω, the function $x^o(t)$, $t \in R$, defined by

$$x^o(t) = \int_o^t \bar{f}_o(\tau, x(\tau), u(\tau))\, d\tau, \quad t \in R,$$

where the integration is made on the interval $[0, t]$ and where 0 and t denote the ν-vectors $0 = (0, \dots, 0)$, $t = (t^1, \dots, t^\nu) \in R$. If $t = (t^1, \dots, t^\nu)$, $t' = (t'^1, \dots, t'^\nu)$ are any two points of R, then $|t^i - t'^i| \le |t - t'|$, $i = 1, \dots, \nu$. Let us consider the finite sequence of points $t_0 = t, t_1, t_2, \dots, t_\nu = t'$ all in R, defined by $t_j = (t^1, \dots, t^{\nu-j}, t'^{\nu-j+1}, \dots, t'^\nu)$, $j = 0, 1, \dots, \nu$, and let r_j denote the interval

$$[0, t^1] \times \dots \times [0, t^{\nu-j-1}] \times [t^{\nu-j}, t'^{\nu-j}] \times [0, t'^{\nu-j+1}] \times \dots \times [0, t'^\nu].$$

Then

$$x^o(t) - x^o(t') = \Sigma_j [x^o(t_j) - x^o(t_{j+1})]$$

where Σ ranges over all $j = 0, 1, \dots, \nu - 1$, and

$$x^o(t_j) - x^o(t_{j+1}) = \int_{r_j} \bar{f}_o(\tau, x(\tau), u(\tau))\, d\tau .$$

Since meas $r_j \leq b^{\nu-1} |t^{\nu-j} - t'^{\nu-j}| \leq b^{\nu-} |t - t'|$, and $\bar{f}_o \leq M_o$, we have

$$|x^o(t) - x^o(t')| \leq \sum_j M_o \text{meas } r_j \leq \nu M_o b^{\nu-1} |t-t'|.$$

In particular $|x^o(t)| = |x^o(t) - x^o(0)| \leq \nu M_o b^\nu$ for every $t \in R$

Thus , the functions $x^o(t)$, relative to the pairs $x(t)$, $u(t)$ of Ω, are equicontinuous, equibounded, equilipschitzian in R. In addition, if for every interval $I \subset R$ we denote by $\Delta_I x$ the usual difference of order ν relative to the 2^ν vertices of I, then

$$|\Delta_I x^o(t)| = \left|\int_I f_o(t, x(t), u(t))dt\right| \leq M_o \text{ meas } I .$$

Thus, if we consider the sequence $x_k(t)$, $u_k(t)$ $t \in R$, we conclude that the functions $x^o_k(t)$ are given by the integrals

$$x^o_k(t) = \int_0^t f_o(t, x_k(t), u_k(t))dt, \ t \in R, \ k = 1, 2, \dots,$$

where the integration is taken in the interval $J = [0, t]$, $0 = (0, \dots, 0)$, $t = (t^1, \dots, t^\nu) \in R$, and these functions are equibounded, equicontinuous, equilipschitzian in R with

$$(47) \qquad |x^o_k(t) - x^o_k(t')| \leq \nu M_o |t-t'| \ , \ |\Delta_I x^o_k| \leq M_o \text{ meas } I,$$

for all $t, t' \in R$, $I \subset R$, By Ascoli's theorem we can perform a further selection and obtain a sequence $[\tilde{x}_{k_r}(t), u_{k_r}(t)]$, $\tilde{x}_{k_r} = (x^o_{k_r}, \dots, x^n_{k_r})$, for which (47) holds, and for which $x^o_{k_r}(t) \rightrightarrows x^o(t)$ uniformly in R as $r \to \infty$, and $x^o(t)$ is continuous in R. In addition $\Delta_I x^o_{k_r} \to \Delta_I x^o$ as $r \to \infty$ for every $I \subset R$. Hence, (47) yields

$$|x^o(t) - x^o(t')| \leq \nu M_o |t - t'|, \quad |\Delta_I x^o| \leq M_o \text{ meas } I,$$

for all $t, t' \in R$, $I \subset R$. Thus $\Delta_I x^o$ is an additive interval function which is absolutely continuous in R. By [14] we know that $\Delta_I x^o$ admits of a derivative $z(t)$ which is L-integrable in R and

$$\Delta_I x^o = \int_I z(t)\, dt$$

for every $I \subset R$. In particular

$$x^o(t) = \Delta_J x^o = \int_o^t z(t) dt, \quad \iota = (t^1, \dots, t^\nu) \in R .$$

Hence $x^o(t)$ admits of a generalized derivative $D^\alpha x^o(t) = z(t)$ a.e. in R, $\alpha = (1, \dots, 1)$, $\alpha = l_o = \nu$, and $D^\alpha x^o(t) = z(t) \leq M_o$ a.e. in R. Also $D^\alpha x^o_k(t) = 0$ in $R - cl\, G$ implies $z(t) = 0$ a.e. in $R - cl\, G$. Finally, the uniform convergence $x^o_{k_r} \to x^o$ and the equibounded-ness of all $x^o_{k_r}$ imply the strong convergence $x^o_{k_r} \to x^o$ in $L_1(R)$. Thus

$$x^o_{k_r}(t) \to x^o(t) \quad \text{strongly in } L_1(R) \text{ as } r \to \infty,$$

$$\int_I D^\alpha x^o_{k_s}(t)dt \to \int_I D^\alpha x^o(t)dt \quad \text{as } r \to \infty, \text{ for } \alpha = (1, \dots, 1), \; I \subset R .$$

We shall now apply Closure Theorem I (and remark at the end of No. 18) to the sequence $\tilde{x}_k(t)$, $\tilde{u}_k(t)$ relative to the integral $J[\tilde{x}, \tilde{u}]$, control space $\tilde{U}(t, x)$, vector function $\tilde{f}(t, x, \tilde{u}) = (\tilde{f}_o, f_{i\alpha})$, $\tilde{f}_o = \tilde{u}^o$, for which the set $\tilde{\tilde{Q}}(t, x) = \tilde{f}(t, x, \tilde{U}(t, x))$ is by hypothesis convex for every $(t, x) \in A$ and satisfies condition (Q) in A. Then there exists a vector function $\tilde{u}(t) = (\tilde{u}^o, u)$, $u(t)$ defined a.e. in G and measurable in G, $\tilde{u}^o(t)$ defined a.e. in R and measurable in R, such that

$$D^{\alpha} x^i(t) = f(t, x(t), u(t)) \text{ a.e. in } G, \alpha \in \{\alpha\}_i, \ i = 1, \dots, n,$$

$$D^{\alpha} x^o(t) = \tilde{u}^o(t) \text{ a.e. in } R, \ \alpha = (1, \dots, 1),$$

hence $\tilde{u}^o(t) = 0$ in $R - cl\, G$, and in addition such that

$$u(t) \in U(t, x(t)) \quad \text{a.e. in } G$$

$$\tilde{u}^o(t) \geq f_o(t, x(t), u(t)) \text{ a.e. in } G,$$

$$J[x, u] = x^o(b, \dots, b).$$

On the other hand,

$$I[x_{k_r}, u_{k_r}] = x^o_{k_r}(b, \dots, b) \longrightarrow i \text{ as } r \longrightarrow \infty,$$

and $x^o_{k_r}(t) \rightarrow x^o(t)$ uniformly in R. Hence

$$J[\tilde{x}, \tilde{u}] = x^o(b, \dots, b) = i.$$

Now the pair $x(t)$, $u(t)$ is certainly admissible and contained in Ω, hence

$$i \leq I[x,u] = \int_G f_o(t, x(t), u(t))dt \leq \int_G \tilde{u}^o(t)dt = \int_R \tilde{u}^o(t)dt = J[\tilde{x}, \tilde{u}] = i,$$

where i is finite. This implies that we must have

$$\tilde{u}^o(t) = f_o(t, x(t), u)) \quad \text{a.e. in } G,$$

and hence

$$i = I[x, u] = \int_G f_o(t, x(t), u(t))dt = J[\tilde{x}, \tilde{u}].$$

This proves that $I[x, u]$ attains the value i in Ω. Theorem VIII is thereby proved.

Corollary: Under the conditions of Theorem VIII with $\tilde{Q}(t, x) = \tilde{f}(t, x, U(t, x))$ convex for every $(t, x) \in A$, then $I[x, u]$ has an absolute minimum in Ω.

Indeed, if $\tilde{Q}$ is convex, so is $\tilde{\tilde{Q}}$. This corollary corresponds to Filippov's existence theorem ,

21. EXAMPLES.

(a) First we consider the example of the minimum of the integral

$$I[x] = \int_G\int (x_\xi^2 + x_\eta^2)\, d\xi\, d\eta$$

in the class Ω of all functions $x(\xi, \eta)$, $(\xi, \eta) \in G$, of the Sobolev's class $W_2^1(G)$, assuming given boundary values ϕ on the boundary $S = \partial G$ of G, and satisfying the constraints $|x_\xi|, |x_\eta| \leq L$. Here G is an open bounded set of the $\xi\eta$-plane and of

Sobolev's class K_1, L is a given constant, and Ω is assumed to be nonempty. The problem can be written in the form of the existence theorem I with $n = 1$, $\mu = 2$, $N = 2$, $m = 2$:

$$I[x, \mu, v] = \iint_G (u^2 + v^2)\, d\xi\, d\eta = \text{minimum},$$

$$x = \phi \text{ on } S, \quad \frac{dx}{d\xi} = u, \frac{\partial x}{\partial \eta} = v, (u, v) \in U = [|u| \le L, |v| \le L].$$

The set $\tilde{\tilde{Q}}$ is the convex set of all (z, u, v) with $|u| \le L, |v| \le L$, $u^2 + v^2 \le z \le 2L^2$. The minimum of $I[x, u]$ in Ω exists by forse by force of Existence Theorem VIII.

(b) Consider the problem discussed in Theorem VIII with $n=1$, $\mu = 2$, $N = 2$, $m = 2$:

$$I[x, u, v] = \iint_G [A(\xi, \eta, x)u^2 + 2B(\xi, \eta, x)\, uv + C(\xi, \eta, x)v^2]\, d\xi d\eta = \text{minimum},$$

$$\frac{\partial x}{\partial \xi} = a(\xi, \eta)u + b(\xi, \eta)\, v,$$

$$\frac{\partial x}{\partial \eta} = c(\xi, \eta)u + d(\xi, \eta)\, v,$$

where G is an open bounded set of the $\xi\eta$-plane of Sobolev's class $W_2^1(G)$, where A, B, C are continuous functions of (ξ, η, x) in $c\ell G \times E_1$ with $A > 0$, $C > 0$, $B^2 < AC$ in $c\ell G \times E_1$, where a, b, c, d are continuous functions of (ξ, η) on $c\ell G$, where we assume boundary conditions $x = \phi$ on $S = \partial G$, ϕ a given continuous function on S, and unilateral constraints $(u, v) \in U = [|u| \leq L, |v| \leq L]$. Let Ω denote the class of all functions $x(\xi,\eta)$ of Sobolev's class $W_2^1(G)$ satisfying the conditions above, and let us assume that Ω is not empty. The function $x(\xi,\eta)$ of Ω are uniformly bounded, say $|x(\xi,\eta)| \leq M$, and we can take for A the set $c\ell G \times [-M, M]$. Let M_0 be the maximum of $Au^2 + 2Buv + Cv^2$ for all $(\xi,\eta) \in c\ell G$, $x \leq M$, $|u| \leq L$, $|v| \leq L$. The set $\tilde{Q}(\xi,\eta, x)$ is the set of all (z^0, z^1, z^2) with $Au^2 + 2Buv + Cv^2 \lesssim z^0 \leq M_0$, $z^1 = au + bv$, $z^2 = cu + dv$, $|u| \leq L$, $v \leq L$, and obviously $\tilde{Q}$ is convex. The minimum of the cost functional in Ω exists by force of the existence theorem VIII.

(c) Consider the problem

$$I[x, u] = \int_0^T \iint_G f_0(t, \xi, \eta, x)\,dt\,d\xi\,d\eta = \text{minimum},$$

$$\frac{\partial x}{\partial t} = a(t,\xi,\eta)\frac{\partial x}{\partial \xi} + b(t,\xi,\eta)\frac{\partial x}{\partial \eta}$$

$$x(o,\xi,\eta) = \phi(\xi,\eta), \quad \left(\frac{\partial x}{\partial \xi}\right)^2 + \left(\frac{\partial x}{\partial \eta}\right)^2 \leq L,$$

where G is a given open set of the $\xi\eta$-plane of Sobolev's class $W_2^1(G)$, where ϕ is a given continuous function on $c\ell G$. If $G_0 = G \times (o, T)$ we shall assume a, b continuous functions on $c\ell G_0 = c\ell G \times [o, T]$. Let Ω be the class $W_2^1(G_0)$ satisfying the conditions above, and we assume Ω not empty. The functions $x(t,\xi,\eta)$ of Ω are uniformly bounded, say $-M \leq x \leq M$, and we assume f_0 to be a continuous function of t, ξ, η, x in $c\ell G_0 \times [-M, M]$ The problem above written in terms of the existence theorem becomes

$$I[x, v, w] = \iiint_{G_0} f_0(t, \xi, \eta, x)\,dt\,d\xi\,d\eta = \text{minimum},$$

$$\frac{\partial x}{\partial \xi} = a(t, \xi, \eta)v + b(t, \xi, \eta)w, \quad x(o, \xi, \eta) = \phi(\xi, \eta),$$

$$\frac{\partial x}{\partial \xi} = v, \quad \frac{\partial x}{\partial \eta} = w, \quad (v, w) \in U = [v^2 + w^2 \leq L],$$

with $n = 1$, $\mu = 3$, $m = 2$, $N = 3$. Then the set $\widetilde{Q}(t, \xi, \eta, x)$ is the set of all (z^0, z^1, z^2, z^3) with $z^0 = c_o = f_o(t, \xi, \eta, x)$, $z^1 = av + bw$, $z^2 = v$, $z^3 = v$, $v^2, w^2 \leq 1$, and hence $\widetilde{Q}_o$ is convex. This existence of the minimum follows from the corollary above.

(d) Let us consider the problem written in terms of the existence theorem with $n = 1$, $\mu = 2$, $m = 2$, $N = 1$:

$$I[x, u, v] = \int_0^1 \int_0^1 f_0(\xi, \eta, x)\,d\xi\,d\eta$$

$$\frac{\partial^2 x}{\partial \xi^2} = a(\xi, \eta)u + b(\xi, \eta)v, \quad u^2 + v^2 \leq 1,$$

$$\int_0^1 \int_0^1 \left(\frac{\partial^2 x}{\partial \eta^2}\right)^2 d\xi\,d\eta \leq L_1, \quad \int_0^1 \int_0^1 \left(\frac{\partial^2 x}{\partial \xi \partial \eta}\right)^2 d\xi\,d\eta \leq L_2,$$

$$x(0, \eta) = \phi(\eta), \quad \partial x(0, \eta)/\partial \xi = \psi(\eta).$$

If G denotes the open square $0 < \xi < 1$, $0 < \eta < 1$, let $R = c\ell\, G$ be the corresponding closed square, a, b to be given continuous functions of (ξ, η) in R, ϕ, ψ continuous functions of η in $[0, 1]$, and L a constant. Let Ω be the class of all functions $x(\xi,\eta)$ of the Sobolev's class $W_2^2(G)$, and assume Ω to be not empty. Then every function $x(\xi,\eta)$ of the class Ω is uniformly bounded in R (together with $\partial x/\partial\xi$, $\partial^2 x/\partial\xi^2$), say $|x| \leq M$. Let $A = R \times [-M, M]$ and assume f_0 to be a continuous function of (ξ, η, x) in A. The boundary conditions above necessarily satisfy conditions (P_1) and (P_2). Also, $\tilde{Q}(\xi,\eta, x)$ is the set of all (z^0, z^1) with $z^0 = c_0 = f_0(\xi,\eta, x)$, $z^1 = au + bv$, $u^2 + v^2 \leq 1$, and $\tilde{Q}$ is then a segment $z^0 = c_0$, $c_1 \leq z^1 \leq c_2$ (depending on ξ,η, x), and $\tilde{Q}$ is convex. The existence of the minimum of the cost functional in Ω follows from the corollary.

REFERENCES

[1] L. Cesari, (a) Semicontinuità del calcolo delle variazioni, Annali Scuola Normale Sup. Pisa, 14, 1964, 389-423: (b) Un teorema di esistenza in problemi di controlli ottimi, Ibid, 1965, 35-78: (c) An existence theorem in problems of optimal control, J. SIAM Control, (A) 3 1965, 7-22; (d) Existence theorems for optimal solutions in Pontryagin and Lagrange problems, J. SIAM Control, (A) 3 1965, 475-498; (e) Existence theorems for generalized and usual optimal solutions in Lagrange problems with unilateral constraints. I and II. Trans. Amer Math. Soc., to appear. (f) Existence theorems for multidimensional problems of optimal control. Conference in Differential Equations, Mayaguez, Puerto Rico, December 1965, to appear.

[2] A. F. Filippov, On certain questions in the theory of optimal control, Vestnik Moskov. Univ. Ser. Mat. Mech. Astr. 2, 1959, 25-32 (Russian). English translation in J. SIAM control (A), 1, 1962, 26-34.

[3] R. V. Gamkrelidze, On sliding optimal regimes, Dokl. Akad. Nauk USSR 143, 1962, 1243-1245 (Russian). English translation in Soviet Math. Doklady 3, 1962, 390-395.

[4] J. R. LaPalm, Existence analysis in problems of optimal control with closed control space and exceptional sets. A. Ph. D. thesis at the University of Michigan, Ann Arbor, Michigan, 1966.

[5] L. Markus and E. B. Lee, Optimal control for nonlinear processes, Arch. Rational Mech. Anal. 8, 1961, 36-58.

[6] E. J. McShane, (a) Curve-space topologies associated with variational problems, Ann. Scuola Norm. Sup. Pisa, (2) 9, 1940, 45-60. (b) Generalized curves, Duke Math. J., 6, 1940, 513-536. (c) Necessary conditions in the generalized-curve problem of the calculues of variations, Ibid., 7, 1940, 1-27 (d) Existence theorems for Bolza problems in the calculus of variations, Ibid, 28-61. (e) A metric in the space of generalized curves, Ann. of Math?, (2) 52, 1950, 328-349.

[7] C. B. Morrey, (a) Multiple integral problêms in the calculus of variations, Univ. of California Publ. in Math. 1, 1943, 1-30; (b) Multiple integrals in the calculus of variations, Grundl, Math. Wiss. Bd. 130, Springer 1966.

[8] M. Nagumo, Uber die gleichmässige Summierbarkeit und ihre Anwendung auf ein Variation problem. Japanese Journ, Math. 6, 1929, 173-182.

[9] A. Plis, Trajectories and quasi trajectories of an orientor field. Bull. Acad. Pol. Sci., 11, 1963, 369-370.

[10] L. S. Pontryagin, Optimal control processes, Uspehi Mat. Nauk 14, 1 (85), 1959, 3-20. English Translation in Amer. Math. Soc. Transl. (2), 1961, 321-339.

[11] L. S. Pontryagin, V. C. Boltyanskii, R. V. Gamkrelidze, and E. F. Mishchenko, The mathematical theory of optimal processes, Gosudarst. Moscow 1961. English translations: Interscience 1962; Pergamon Press 1964.

[12] E. H. Rothe, An existence theorem in the calculus of variations based on Sobolev's imbedding theorems. Archive Rat. Mech. Anal. 21, 1966, 151-162.

[13] E. Roxin, The existence of optimal controls, Mich. Math. J., 9, 1962, 109-119.

[14] S. Saks, Theory of integral, Monogr. Matem, 1937.

[15] S. L. Sobolev, Applications of functional analysis in mathematical physics, Amer, Math. Soc. Translation, Providence, R.I., 1963.

[16] L. Tonelli, (a) Sugli integrali del calcolo delle variazioni in forma ordinaria, Annali Scuola Normale Sup. Pisa, (2) 3, 1934, 401-450. Opere Scelte, Cremonese, Roma 1962, 3, 192-254; (b) Un teorema di calcolo delle variazioni, Rend. Accad. Lincei, 15 1932, 417-423. Opere Scelte, 3 84-91; (c) Fondamenti di calcolo delle variazioni, Zanichelli, Bologna 1921-23.

[17] L. Turner, The direct method in the calculus of variations, A. Ph.D. thesis at Purdue University, Lafayette, Indiana, 1957.

[18] A. Turowicz, (a) Sur les trajectoires et quasi trajectoires des systèmes de commande non linéaires. Bull. Acad. Pol. Sci. 10, 1962, 529-531; (b) Sur les zones d'émission des trajectoirès et des quasi trajectoires des systèmes de commande non linéaires. Ibid., 11, 1963, 47-50.

[19] J. Warga, Relaxed, variational problems, J. Math. Anal. Appl. 4 1962, 111-128.

[20] T. Wazewski, (a) Sur une généralization de la notion des solutions d'une équation au contingent, Bull. Acad. Polon. Sci. 10, 1962 11-15; (b) Sur les systèmes de commande non linéaires dont le contredomaine de command n'est pas forcément convexe, Ibid., 10, 1962, 17-21.

[21] L. C. Young, Generalized curves and the calculus of variations, C. R. Soc. Sci. Varsovie, (3) 30, 1937, 212-234.

CENTRO INTERNAZIONALE MATEMATICO ESTIVO
(C. I. M. E.)

Hubert HALKIN

OPTIMAL CONTROL AS PROGRAMMING IN INFINITE DIMENSIONAL SPACES

Corso tenuto a Bressanone dal 10 al 18 giugno
1966

OPTIMAL CONTROL AS PROGRAMMING IN INFINITE DIMENSIONAL SPACES*

by

Hubert HALKIN

(University of California, San Diego)

Introduction. The terms "Programming" and "Mathematical Programming" refer here to the usual constrained maximization problem of the type : given a function $\varphi(x) = (\varphi_1(x), \ldots, \varphi_k(x))$ from E^n (the Euclidean n-dimensional space) into E^k, find a point $\hat{x} \in E^n$ such that $\varphi_1(\hat{x})$ is maximized subject to equality and/or inequality constraints on $\varphi_2(\hat{x}), \ldots, \varphi_k(\hat{x})$. This problem has received considerable attention and has led to interesting results ranging from the initial work of Lagrange to the more recent studies of Kuhn and Tucker. Calculus of Variations and Optimal Control are also concerned with constrained maximization problems but over given sets of continuous curves instead of finite dimensional Euclidean spaces as in Programming. Programming has always extended a strong influence on Calculus of Variations and one of the motivating forces behind the creation of Functional Analysis was to build a bridge between these two fields. The methods of Functional Analysis have

* This research was sponsored by the Air Force Office of Scientific Research, Office of Aerospace Research, United States Air Force, under AFOSR Grant No. 1039-66 . The present paper is the write-up of lectures given by the author at the Centro Internazionale Matematico Estivo (C. I. M. E.), Bressanone, Italy from June 10 to June 18, 1966 (Session on Calculus of Variations, Classical and Modern).

been used to derive Euler-Lagrange equation but we do not know any previous successful attempts to apply those methods to the derivation of the Weierstrass-E test and the Multiplier Rule for the problem of Bolza. This is indeed the purpose of the present paper. In Section I we shall study a mathematical programming problem in infinite dimensional spaces. In Section II we prove that the standard optimal control problem (a generalization of the problem of Bolza) can be casted into a problem of the type studied in Section I, and by applying to this problem the results of Section I we obtain a generalization of the Maximum Principle of Pontryagin which is itself a generalization of the classical Weierstrass-E test and of the Multiplier Rule for the problem of Bolza (including the abnormal case).

The same methods can be applied to more difficult problems, Halkin-Neustadt [6], but in the present paper we prefer simplicity to utmost generality since these "simple" results contain already the standard necessary conditions of classical calculus of variations and modern optimal control theory. The reader will find in Gamkrelidze [1] and [2], Halkin [3] and [4], Halkin-Neustadt [6], and Neustadt [7] and [8] a proper background for the present paper.

Functional analysis methods and the theory of linear and convex programming over infinite dimensional spaces have been applied previously to study linear control problems with linear or convex constraints whereas no such linearity or convexity assumptions are made concerning the control problems studied in the present paper.

H. Halkin

Section I. A Mathematical Programming Problem in Infinite Dimensional Spaces.

We are given a normed linear space X , a subset L of X and a mapping $\varphi(x) = (\varphi_1(x), \varphi_2(x), \ldots, \varphi_k(x))$ from X into E^k. The problem is to find an $\hat{x} \in L$ such that $\varphi_1(\hat{x})$ is maximized over L subject to the constraints $\varphi_i(\hat{x}) = 0$, $i = 2, 3, \ldots, k$. We assume that such $\hat{x}$ exists . The purpose of the present section is to derive some necessary conditions for $\hat{x}$ to be a solution of this mathematical programming problem. In order to obtain interesting necessary conditions we have to add some structure to our problem. We do that by making two assumptions. The first assumption is very "natural" whereas the second, which is closely related to Kuhn-Tucker constraint qualifications, appears at first to be very artificial, and will be justified only in Section II.

Assumption I. The mapping φ is Fréchet differentiable at $\hat{x}$. We shall denote by $h = (h_1, h_2, \ldots, h_k)$ the continuous linear mapping from X into E^k such that

$$\lim_{|x - \hat{x}| \to 0} \frac{|\varphi(x) - \varphi(\hat{x}) - h(x - \hat{x})|}{|x - \hat{x}|} = 0$$

Assumption II. There exists a set M which is an approximation of L around $\hat{x}$ in the following sense: the closure of the set M is convex and for any set $\{x_1, \ldots, x_k\} \subset M$ such that the vectors

$x_1 - \hat{x}, \ldots, x_k - \hat{x}$ are linearly independent there exists a mapping ζ from $\text{co}\{\hat{x}, x_1, \ldots, x_k\}$ (the convex hull of the set $\{\hat{x}, x_1, \ldots, x_k\}$) into L such that

(α) for any $\delta \in [0, 1]$ the mapping $\zeta(\hat{x} + \delta(x - \hat{x}))$ is continuous with respect to x over $\text{co}\{x_1, \ldots, x_k\}$

$$(\beta) \lim_{\substack{\delta \to 0 \\ \delta > 0}} \frac{\zeta(\hat{x} + \delta(x - \hat{x})) - \hat{x}}{\delta} = x - \hat{x}, \text{ uniformly}$$

over $\text{co}\{x_1, \ldots, x_k\}$.

The central result of this section is then:

<u>Theorem I.</u> (Abstract Maximum Principle). If $\hat{x}$ is an optimal solution and if Assumptions I and II hold then there exist real numbers $\lambda_1, \lambda_2, \ldots, \lambda_k$ such that

$$(\alpha) \sum_{i=1}^{k} |\lambda_i| > 0$$

$$(\beta) \lambda_1 \geq 0$$

$$(\gamma) \sum_{i=1}^{k} \lambda_i h_1(x) \leq \sum_{i=1}^{k} \lambda_i h_i(\hat{x}) \text{ for all } x \in M$$

<u>Proof of Theorem I.</u>

There is no loss of generality by assuming* that $\hat{x} = 0$ and that

---

* This can always be achieved by introducing a new independent variable y by the relation $y = x - \hat{x}$ and a new functional ψ by the relation $\psi(y) = \psi\ x - \hat{x}) = \varphi(x) - \varphi(\hat{x})$.

$\varphi(\hat{x}) = 0$. Let $K = \{(\alpha_1, \ldots, \alpha_k) : \alpha_1 > 0, \alpha_i = 0 \text{ for } i = 2, \ldots, k\}$. We have $K \cap \varphi(L) = \emptyset$. We need to prove that the sets K and h(M) are separated. We shall assume that the sets K and h(M) are not separated and show that this leads to $K \cap \varphi(L) \neq \emptyset$. The sets K and $\bar{h}(M)$, the closure of h(M), are convex. If the sets K and h(M) are not separated then there exists a set $\{x_1, \ldots, x_k\} \subset M$, an element $\alpha^* \in K$ and an $\varepsilon \in (0,1)$ such that $N(\alpha^*, \varepsilon)$, the ε neighborhood of α^*, is a subset of $co\{0, h(x_1), \ldots, h(x_k)\}$. We can always assume that $h_1(x_1) = h_1(x_2) = \ldots = h_1(x_k) > 0$ and, by a change of variables, that $h_1(x_1) = h_1(x_2) = \ldots = h_1(x_k) = +1$. The set $co\{0, h(x_1), \ldots, h(x_k)\}$ will be denoted by H. We shall now define a mapping $\Psi : H \to E^k$ as follows. $\Psi(y) = \varphi(\zeta(x))$ where y is the element of H which has with respect to $\{0, h(x_1), \ldots, h(x_k)\}$ the same barycentric coordinates as x has with respect to $\{0, x_1, \ldots, x_k\}$. From the properties of the functions φ and ζ we know that there exists an $\eta > 0$ such that $|\Psi(y) - y| \leq \eta\varepsilon$ for all $y \in \eta H$ where ηH is the set $\{\eta y : y \in H\}$. Let π be a projection from E^k into E^{k-1} defined by $\pi(\alpha_1, \alpha_2, \ldots, \alpha_k) = (\alpha_2, \ldots, \alpha_k)$. We have then $\pi(\alpha^*) = 0$ and $N(0, \varepsilon) \subset \pi(H)$, where $N(0, \varepsilon)$ is the ε neighborhood of 0 in E^{k-1}. We shall define a mapping $\Phi : \eta\pi(H) \to E^{k-1}$ as follows: $\Phi(y_2, \ldots, y_k) = (y_2, \ldots, y_k) - \pi(\Psi(\eta, y_2, \ldots, y_k))$. The mapping Φ is continuous and $\Phi(\eta\pi(H)) \subset \eta\pi(H)$ since we have $|\Phi(y_2, \ldots, y_k)| = |\pi(\Psi(\eta, y_2, \ldots, y_k)) - (y_2, \ldots, y_k)| \leq |\Psi(\eta, y_2, \ldots, y_k) - (\eta, y_2, \ldots, y_k)| \leq \eta\varepsilon$ and $N(0, \eta\varepsilon) \subset \eta\pi(H)$.

From Brouwer's Fixed Point Theorem there exists a point $(y_2^*, \ldots, y_k^*) \in \eta\pi(H)$ such that $\Phi(y_2^*, \ldots, y_k^*) = (y_2^*, \ldots, y_k^*)$, i.e. such that $\pi(\Psi(\eta, y_2^*, \ldots, y_k^*)) = 0$. We have then $\Psi(\eta, y_2^*, \ldots, y_k^*) = \eta^*$ for some $\eta^* > 0$, (since $|\psi(y) - y| < \eta$ for all $y \in \eta H$) i.e. $\psi(H) \cap K \neq \phi$ and $\varphi(L) \cap K \neq \phi$. This completes the proof of Theorem I.

Section II . Optimal Control Problem .

In this section the normed linear space X is particularized as the space of all absolutely continuous functions from $[0, 1]$ into E^n. We are given a set F of functions $f(z, t)$ from $E^n \times [0, 1]$ into E^n and, as in Section I, a mapping $\varphi = (\varphi_1, \ldots, \varphi_k)$ from X into E^k. The matrix of first partial derivatives of f with respect to z will be denoted by Df. We shall assume that for each $f \in F$ the matrix Df exists and that f and Df are continuous in z and piecewise continuous in t. Let L be the set of all $x \in X$ such that for some $f \in F$ we have $\frac{d}{dt} x(t) = f(x(t), t)$ for a. e $t \in [0, 1]$. The problem is again to find an element $\hat{x} \in L$ which maximizes $\varphi_1(\hat{x})$ over L subject to the constraints $\varphi_1(\hat{x}) = 0, i = 2, \ldots, k$. As in Section I we assume that such $\hat{x}$ exists. In order to apply Theorem I we have to make Assumptions I and II. Assumption I is very natural indeed:in Calculus of Variations and Optimal Control Theory the mapping φ is usually made up valuations at time $t = 0$ and $t = 1$, i. e. one requires the initial and terminal points to belong to some smooth manifolds and φ_1, the payoff functional, is a valuation at time $t = 1$. Assumption II needs the somewhat longer justification which kollows. We shall denote by $\hat{f}$ the element of F such that $\frac{d}{dt}\hat{x}(t) = f(\hat{x}(t), t)$ for a. e. $t \in [0, 1]$. The set M considered in Assumption II will be the set of all solutions x of the linearized variational equation $\frac{d}{dt} x(t) = f(\hat{x}(t), t) + Df(\hat{x}(t), t)(x(t) - \hat{x}(t))$ for a. e. $t \in [0, 1]$, corresponding to all $f \in F$. At this point it is far from obvious that Assumption II is satisfied for the usual optimal control problem and we shall delay the proof of that fact to the end of the present section.

Accepting Assumptions I and II we are now entitled to apply Theorem I to the optimal control problem defined in this section. The integral form of Pontryagin Maximum Principle, stated below, is a particular instance of Theorem I in the case of a functional which has the particular form

$$\varphi(x) = (g_1(x(1)), g_2(x(1)), \ldots, g_\ell(x(1)), g_{\ell+1}(x(0)), \ldots, g_k(x(0)))$$

for some integer ℓ with $1 \leq \ell < k$ and some smooth function $g=(g_1, \ldots, g_k)$ from E^n into E^k.

<u>Theorem II</u> . <u>(Integral form of Pontryagin Maximum Principle</u>).

If φ has the particular form indicated above, if Assumptions I and II hold, and if $\hat{x}$ is an optimal solution, then there exists a $p \in X$ such that

(i) $p(t) = -(Df(x(t),t))^T p(t)$ for a. e. $t \in [0,1]$ where the superscript T indicates transposition of a matrix

(ii) $\int_0^1 p(t) \cdot \hat{f}(\hat{x}(t),t)dt \geq \int_0^1 p(t) \cdot f(x(t),t)dt$

for all $f \in F$

(iii) there exists real numbers $\alpha_1, \ldots, \alpha_k$ such that

(1) $\alpha_1 \geq 0$ and $\sum_{i=1}^{k} |\alpha_i| > 0$

$$(2)\quad p(1) = \sum_{i=1}^{\ell} \alpha_i \nabla g_i(\hat{x}(1))$$

$$(3)\quad p(0) = \sum_{i=\ell+1}^{k} \alpha_i \nabla g_i(\hat{x}(0))$$

where ∇g denotes the gradient of g.

Theorem II is an immediate consequence of Theorem I . The proof requires elementary algebraic manipulations only.

Let us return to the discussion of Assumtion II . In Theorem III we shall prove that Assumption II always holds if the class F has a further property which we call <u>convexity - under - switching</u> and which means that if f_1 and $f_2 \in F$ and $\tau \in (0,1)$ then the function f defined by

$$f(z,t) = f_1(z,t) \quad \text{for} \quad t \in [0,\tau]$$
$$= f_2(z,t) \quad \text{for} \quad t \in (\tau, 1]$$

belong also to the class F. The property of convexity-under-switching is always satisfied in classical calculus of variations and in most optimal control problems. In Gamkrelidze [1] and [2], Halkin [4] and Neustadt [7] and [8] various classes F, besides the convex-under-switching class, have been shown to satisfy Assumption II. Unfortunately the definitions of these general classes are very clumsy and when one tries to streamline these definitions, one arrives always at Assumption II itself.

<u>Theorem III</u>. Assumption II holds for a class F which is convex-under-switching.

The two main ingredients of the proof of Theorem III, given at the end of this section, are (i) standard methods from the theory of ordinary differential equations, in particular Gronwall inequality, and (ii) some results related to Lyapounov's Theorem (see Halkin [5]). We shall first give two Lemmata concerning those latter results.*

<u>Lemma I</u>. If u is an integrable function from [0,1] into E^m and $\mathcal{Q}$ is the set of all subset of [0,1] which are the union of a finite number of intervals, then for any $\varepsilon > 0$ there exists a nest $\{N(t) : t \in [0,1]\} \subset \mathcal{Q}$ such that

(i) $\mu(N(t)) = t$ for all $t \in [0,1]$

(ii) $N(t_1) \subset N(t_2)$ if $t_1 \leq t_2$

(iii) $\left| \int_{N(t) \cap [0,\tau]} u \, d\mu - t \int_0^\tau u \, d\mu \right| \leq \varepsilon$

for all $t \in [0,1]$ and all $\tau \in [0,1)$.

<u>Proof of Lemma I</u>. We first remark that the proof is obvious if the function u is piecewise constant. We know that there is a

---

* The results given here could also be obtained as a corollary to Proposition I of Halkin [5], page 273. However the proof of these earlier results can be drastically simplified as is shown in the present proof of Lemma I. This proof is due to L. W. Neustadt.

sequence $u_1, u_2, \ldots$ of piecewise constant functions from $[0,1]$ into E^m such that $\lim_{i\to\infty} \int_0^1 |u - u_i| \, d\mu = 0$. A positive number ε is given. Let j be a natural number such that $\int_0^1 |u - u_j| \, d\mu \leq \frac{\varepsilon}{4}$ and let $\{N(t): t \in [0,1]\}$ be the nest satisfying conditions (i), (ii) and (iii) for the piecewise constant function u_j and the positive number $\frac{\varepsilon}{2}$. It is a trivial matter to verify that the nest $\{N(t): t \in [0,1]\}$ satisfies conditions (i), (ii) and (iii) for the function u and the positive number ε.

Lemma II. Let $u_1, u_2, \ldots, u_k$ be integrable functions from $[0,1]$ into E^n, let $\Lambda_k = \{\alpha = (\alpha_0, \ldots, \alpha_k) : \alpha_i \geq 0, \sum_{i=0}^k \alpha_i = 1\}$ and let $\eta_k = \{A = (A_0, \ldots, A_k) : A_i \cap A_j = \phi \text{ if } i \neq j,\ A_i \in \mathcal{Q}, \bigcup_{i=0}^k A_i = [0,1]\}$. On the set η_k we shall define the metric $d(A', A'') = \sum_{i=0}^k (A'_i \,\Delta\, A''_i)$ where $A'_i \,\Delta\, A''_i$ denotes the symmetric difference of the sets A'_i and A''_i. Then for any $\varepsilon > 0$ there is a continuous mapping A from Λ_k into η_k such that for all $\alpha \in \Lambda_k$ we have

(i) $\mu(A_i(\alpha)) = \alpha_i$

(ii) $\left| \sum_{i=1}^k \alpha_i \int_0^\tau u_i \, d\mu - \sum_{i=1}^k \int_{A_i(\alpha) \cap [0,\tau)} f_i \, d\mu \right| \leq \varepsilon$

for all $\tau \in [0,1]$.

Proof of Lemma II. Let $\varepsilon > 0$. Let $u = (u_0, \ldots, u_k)$ i.e. u is an integrable function from $[0,1]$ into $E^{n(k+1)}$. Let $\{N(t): t \in [0,1]\}$

be the nest given by Lemma I for the function u and the positive number $\frac{\varepsilon}{k+1}$. Let A be the mapping from Λ_k into η_k defined by $A_0(\alpha) = N(\alpha_0)$

and $A_{i+1}(\alpha) = D\left(\sum_{j=0}^{i+1} \alpha_j\right) \sim D\left(\sum_{j=0}^{i} \alpha_j\right)$ for $i = 0, \ldots, k-1$.

It is a trivial matter to verify that this mapping A satisfies all the conclusions of Lemma II.

Proof of Theorem III. There is no loss of generality by assuming* that $\hat{x} = 0$, $\varphi(\hat{x}) = 0$, $\hat{f}(\hat{x}(t), t) = 0$ and $D\hat{f}(\hat{x}(t), t) = 0$ for all $t \in [0,1]$. Let $\{x_1, \ldots, x_k\}$ be a set of k linearly independent elements of M. We have to construct a mapping ξ from $\mathrm{co}\{0, x_1, \ldots x_k\}$ into L satisfying conditions (α) and (β) of Assumption II. Given any $\delta \in [0,1]$ we shall actually define below $\xi(\delta x) \in L$ for all $x \in \mathrm{co}\{x_1, \ldots, x_k\}$. Let $f_1, \ldots, f_k$ be the elements of F such that for $i=1, \ldots, k$ we have $\frac{d}{dt} x_i(t) = f_i(0,t)$ for a.e. $t \in [0,1]$. Let $u_1, \ldots, u_k$ be functions from $[0,1]$ into E^n defined by $u_i(t) = f_i(0,t)$ for $i=1, \ldots, k$ and $t \in [0;1]$.

---

* This can always be achieved by introducing a new state variable y by the time varying transformation

$$y(t) = G(t)\,(x(t) - \hat{x}(t))$$

where $G(t)$ is the absolutely continuous solution of the matrix differential system

$$G(1) = I$$

$$\frac{d}{dt} G(t) = -G(t) D\hat{f}(\hat{x}(t), t) \quad \text{a.e. } t \in [0,1]$$

and by defining a new functional ψ by the relation $\psi(y) = \varphi(x - \hat{x}) = \varphi(x) - \varphi(\hat{x})$.

Let A be the mapping given by Lemma II for the positive number $\varepsilon = \delta^2$.

We define now $\alpha(\delta x) \in \Lambda_k$ for all $x \in co\{x_1, \ldots, x_k\}$ as follows: $\alpha(\delta x)$ has the same barycentric coordinates with respect to Λ_k as δx has with respect to $\{0, x_1, \ldots, x_k\}$. For every $x \in co\{x_1, \ldots, x_k\}$ we define $f_{\delta x} = \sum_{i=0}^{k} \chi(A_i(\alpha(\delta x))) f_i$ where χ indicates the characteristic function of a set. We have then $f_{\delta x} \in F$. Finally we define $\zeta(\delta x)$ as the absolutely continuous solution of the differential system

$$z(0) = \delta x(0)$$
$$\dot{z} = f_{\delta x}(z,t) \text{ for a. e. } t \in [0,1],$$

With the help of the theory of ordinary differential equations (in particular, Gronwall inequality) one can prove that there exists a $\delta^* \in [0,1]$ such that the definition of $\zeta(\delta x)$ given above makes sense for $\delta \in [0, \delta^*]$ (if $\delta \in (\delta^*, 1]$ we let $\zeta(\delta x) = 0$) and we can prove that this function ζ satisfies conditions (α) and (β) of Assumption II.

Acknowledgment. The author's papers can be divided into two categories: those about which Lucien W. Neustadt made valuable comments before publication and those about which he made valuable comments after publication. The present paper belongs to the first category.

REFERENCES

[1] Gamkrelidze, R. V., On the Theory of the First Variation, Dok. Akad. Nauk., 161, 1965, 23-26. English translation in Soviet Mathematics, 6, 1965, 345-348.

[2] Gamkrelidze, R. V., On Some Extremal Problems in the Theory of Differential Equations with Applications to the Theory of Optimal Control, SIAM J. of Control, 3, 1965, 106-128.

[3] Halkin, H. An abstract framework for the theory of process optimization Bull. Amer. Math. Soc., 72(1966), pp. 677-678.

[4] Halkin H., Finite Convexity in Infinite Dimensional Spaces, Proceedings of the Colloquium on Convexity, Copenhagen, 1966, W. Fenchel, ed., pp. 126-131.

[5] Halkin H., Some further generalizations of a theorem of Lyapounov, Arch. Rat. Mech. An. 17, 1964 pp. 272-277.

[6] Halkin H. and L. W. Neustadt, General Necessary Conditions for Optimization Problems, to appear in Proc. Nat. Acad. Sci.

[7] Neustadt L. W., An abstract variational theory with applications to a broad class of optimization problem I: General Theory, to appear in SIAM J. of Control 4 (1966). Available as report of the Electronic Sciences Laboratory, University of Southern California, Los Angeles, California.

[8] Neustadt L. W. An abstract variational theory with applications to a broad class of optimization problem II: Applications, to appear. Available as report of the Electronic Sciences Laboratory, University of Southern California, Los Angeles, California.

CENTRO INTERNAZIONALE MATEMATICO ESTIVO

(C. I. M. E.)

Czeslaw OLECH

"THE RANGE OF INTEGRALS OF A CERTAIN CLASS OF VECTOR-VALUED FUNCTIONS"

Corso tenuto a Bressanone dal 10 al 18 giugno 1966

The range of integrals of a certain class of vector-valued functions

by

Czelaw Olech(*)

(Inst. of Math., PAN, Krakow)

Introduction. Let J be a compact interval of R^1, say $[0, 1]$. Consider the metric space (M, ρ), where $M = M_{R^n}(J)$ is the space of Lebesgue measurable functions of J into R^n and the metric function ρ is given by

$$(0, 1) \qquad \rho(f, g) = \inf_{\alpha > 0} \left(\alpha + \mu(s: |f(s) - g(s)| > \alpha)\right), \quad f, g \in M,$$

where $|\ |$ stands for a norm in R^n and μ denotes the Lebesgue measure in R^1. We identify two functions of M if they differ on a set of measure zero.

Definition 1. Let $S \subset M$. We say that $u \in M$ is piece-wise in S or belongs piece-wise to S iff there exists a finite number of functions $u^1, \ldots, u^k \in S$ and a decomposition of J into k mutually disjoint subintervals $J_1, \ldots, J_k$ such that $u(t) = u^i(t)$ if $t \in J_i$ for each $i = 1, \ldots, k$. We denote by P(S) the set of all functions which are piece-wise in S.

The purpose of the present note is to study the range of integrals of a class K of vector-valued functions, which satisfies the following three conditions.

(i) K is closed in (M, ρ).

(*)During the preparation of this paper the author was visiting Istituto Matematico "Ulisse Dini", Università di Firenze invited by Consiglio Nazionale delle Ricerche.

(ii) $\left|\int_J u\,dt\right| \leq m < +\infty$ for each $u \in K$, where m is a constant

(iii) $P(K) = K$; that is each function which is piece-wise in K also belongs to K.

These conditions concerning K will be assumed throughout the paper without further mentioning.

Note that no convexity assumptions are imposed on K. Also we do not assume that elements of K are bounded by an integrable function but we assume instead a weaker condition (ii).

The results we are going to present here represent a generalization of the principal results of a previous paper [1]. The difference is that in [1] we were concerned with a set-valued function G of J into the space of compact subsets of R^n which is measurable (cf. [2]). The class K was there defined as the set of all measurable $u: J \longrightarrow R^n$ such that $u(t) \in G(t)$ a.e. in J. It is clear that the so defined class K satisfies conditions (i) and (iii). A stronger version of condition (ii) was additionally assumed in [1].

The results of this paper have some worthwhile applications to the time optimal control problems which are not discussed here. For more details we refer the reader to the paper [1], where also a more complete list of references can be found.

<u>Lexicographic order.</u> Let $\xi = (\xi_1, \ldots, \xi_n)$ be an orthonormal basis of R^n. By Ξ we denote the set of all such possible bases. Let $x, y \in R^n$ and x_i, y_i denote the i-th coordinate of x and y respectively with respect to the basis ξ. Then we write $x \underset{\xi}{<} y$ iff there exists $k \leq n$ such that $x_k < y_k$ while $x_i = y_i$ for $i = 1, \ldots k-1$ if $k > 1$.

Let $A \subset R^n$. By $e(A, \xi) \in A$ we denote the lexicographical maximum of A; that is a point of A such that for each $a \in A$, $a \neq (A, \xi)$ we have $a \underset{\xi}{<} e(A, \xi)$. If A is compact then $e(A, \xi)$ exists and is unique for each $\xi \in \Xi$. Moreover (cf. [3]), the set $E = \{e(A, \xi) : \xi \in \Xi\}$ is the profile of the convex hull of A. Let us recall that the profile of a convex set is the set of all extreme points and a point of a convex set is called extreme if it is not an interior point of any interval contained in the set. The profile of a convex set B is denoted by $\ddot{B}$.

If $u, v \in M$, we shall write $u \underset{\xi}{<} v$ iff for almost all t in J either $u(t) = v(t)$ or $u(t) \underset{\xi}{<} v(t)$ and the second possibility holds on a set of positive measure. If $S \subset M$, we say that $e(S, \xi) \in S$ is the lexicographical maximum of S with respect to ξ if for each $u \in S$, $u \neq e(S, \xi)$ we have the inequality $u \underset{\xi}{<} e(S, \xi)$. Note that if $e(S, \xi)$ exists then it is unique. We have the following

Theorem 1. If K satisfies (i), (ii) and (iii) then for each $\xi \in \Xi$ the lexicographical maximum $e(K, \xi)$ of K exists.

Range of integrals. For the sake of brevity we denote by

$$I(u) = \int_J u\,dt = \int_0^1 u\,dt \quad \text{and} \quad I_t(u) = \int_0^t u\,dt, \quad t \leq 1 .$$

$$I(K) = \{I(u) : u \in K\} \quad \text{for any subset } K \subset M .$$

It is easy to see that if $u \underset{\xi}{<} v$ then also $I(u) \underset{\xi}{<} I(v)$. Therefore by Theorem 1 we have

$$I(e(K, \xi)) = e(I(K), \xi) \quad \text{for each } \xi \in \Xi. \tag{2.1}$$

Note also the implication

(2.2) $\quad$ if $I(u) = I(e(K, \xi))$ and $u \in K$ then $u = e(K, \xi)$.

Theorem 2. The set $E = \{x: x = I(e(K, \xi)),\ \xi \in \Xi\}$ is the profile of a convex and compact set . More precisely there exists a convex and compact set $B \subset R^n$ such that

(2.3) $\quad E = \ddot{B}$ and $e(B, \xi) = I(e(K, \xi))$ for each $\xi \in \Xi$.

From (2.1) and Theorem 1 we have the following

Theorem 3. If B is the set in Theorem 2 then $I(K) \subset B$.

It is obvious that if J is replaced by $[0, t]$, $0 < t \leq 1$, and I by I_t then all results stated up to now hold true. We will consider the corresponding sets $E(t)$ and $B(t)$. Hence E and B in Theorem 2 are in the present notations $E(1)$ and $B(1)$ respectively . We have

(2.4) $\quad E(t) = \ddot{B}(t)$ and $I_t(K) \subset B(t)$.

Theorem 4. The set valued function B(t) is continuous in t in the sense of Hausforff distance ; that is

(2.5) $$\max_{a \in B(s),\, b \in B(t)} (r(a, B(t)), r(b, B(s))) \longrightarrow 0 \quad \text{as} \quad |s-t| \longrightarrow 0$$

where $r(a, B)$ denotes the distance of a point a from set B .

Finally we have

Theorem 5. For each $b \in B$, where B is given by Theorem 2, there exist two sequences $\xi^1, \dots \xi^k \in \Xi$ and $t_o = 0 < t_1 < \dots t_k = 1$,

$k \leqslant n+1$ such that for the function

(2.6) $$u = e(K, \xi^i) \quad \text{if} \quad t_{i-1} \leqslant t < t_i , \qquad i = 1, \dots, k ,$$

we have

(2.7) $$b = I(u) .$$

Moreover, if $b \in \ddot{B}$ then k is equal to 1 and the u satisfying (2.7) is unique while if $b \notin \ddot{B}$ then there are at least two different u satisfying (2.7) and defined by (2.6).

From (iii) and Theorem 1 each u defined by (2.6) belongs to K. Hence Theorem 5 implies that $I(K) \supset B$, which together with Theorem 3 gives equality. Thus we have

Theorem 6. The range of integrals over the class K satisfying (i), (ii) and (iii) is compact and convex.

Denote by e(K) the subset of K consisting of all $e(K, \xi)$, $\xi \in \Xi$. Points of e(K) will be called extremal elements of K. Another consequence of Theorem 5 is the following

Theorem 7. A function $u \in K$ is an extremal element of K iff the following implication holds:

if $v \in K$ and $I(v) = I(u)$ then $u = v$.

Consider now functions which are piecewise in e(K), which we will call piece-wise extremal. We denote for brevity $K_o = P(e(K))$. Manifestly $P(K_o) = K_o$ and by Theorem 5 $I(K_o) = I(K)$

Theorem 8. If $K_1 \subset K$ satisfies the two properties

(2.8) $$P(K_1) = K_1 \quad \text{and} \quad I(K_1) = I(K)$$

then $K_1 \supset K_o$. Hence K_o is the smallest subclass of K having the two properties (2.8).

Concerning the proofs. As has been pointed out the difference between the results presented here and those in [1] lies in the fact that the class K we consider here is more general. However, as far as proofs are concerned only Theorem 1 and Theorem 2 require a new proof and the proofs of the remaining theorems can be traced in [1]. Detailed proofs of the present results will be published elsewhere.

REFERENCES

[1] Olech, C., Extremal solutions of a control system, J. Diff. Eq., 2(1966), 74-101.

[2] Olech, C., A note concerning set-valued measurable functions, Bull. Acad. Polon. Sci. Ser. Sci. Math. Astron. Phys. 13(1965) 317-321.

[3] Olech, C., A note concerning extremal points of a convex set. Bull. Acad. Polon. Sci., Ser. Sci. Math. Astron. Phys. 13(1965), 347-351.

CENTRO INTERNAZIONALE MATEMATICO ESTIVO

(C. I. M. E.)

E. H. ROTHE

WEAK TOPOLOGY AND CALCULUS OF VARIATIONS

Corso tenuto a Bressanone dal 10 al 18 giugno

1966

WEAK TOPOLOGY AND CALCULUS OF VARIATIONS

by

E. H. Rothe

1. <u>Introduction.</u> It might be said that the use of weak compactness in the calculus of variations is implicitly contained in the classical method of proving existence theorems by selecting weakly convergent subsequences. In recent years the use of weak topology in the calculus of variations has become a good deal more explicit and systematic. It is based on the following facts :

I. In any topological space E a real valued function f defined on a compact subset A takes a (absolute) minimum in some point of A if it is lower semi-continuous, and also a maximum if it is continuous,

II. The closed unit ball in a reflexive Banach space is compact in the weak topology.

III. Many of the function spaces used in the calculus of variations are reflexive Banach spaces, e.g. the L_p spaces for $p > 1$, the spaces of Calkin and Morrey, and the related spaces of Sobolev.

IV. Convexity of a function f (together with a certain boundedness condition) guarantees "weak" lower semi-continuity, while for weak continuity the "compactness" of the Gateaux derivative $\delta f(x; h)$ is sufficient.

The object of the present, for a good deal expository, paper is to discuss the points above and to give some applications.

In section 2 the facts stated in I are proved; this sectinn

also contains a theorem on convex functions used later on.

Section 3 reviews the notions of weak topology and reflexivity in a Banach space E. Here a word on the terminology is in order: the word "weakly" will always refer to the weak topology; thus "the real valued f is weakly continuous at $x = x_0$" means that f is continuous at x_0 in the weak topology of E. This does not quite agree with the traditional terminology according to which the above statement in quotes means:

$$\lim_{n \to \infty} f(x_n) = f(x_0) \quad \text{for every sequence } \{x_n\} \ (n = 1, 2, \ldots)$$

which has the property that $\lim_{n \to \infty} \ell(x_n) = \ell(x_0)$ for every bounded linear functional ℓ.

In the present paper the latter property will be referred to as weak sequential continuity. While in general weak continuity and weak sequential continuity are not the same, it will be shown that for the important class of bounded sets in reflexive and separable Banach spaces the two notions agree.

Section 4 contains existence theorems for extrema and the theorem on convex functions mentioned in IV above.

In section 5 the connection between weak continuity (and a closely related property) of f and properties of the Gâteaux differential $\delta f(x; h)$ is discussed.

Section 6 deals with applications to Sobolev spaces and to the calculus of variations, the decisive tool being the inequality (6.3).

2. General theorems on lower semi-continuous and on convex functions. Let E be a topological space [1]. A real valued function f on E is a map $E \rightarrow R$ where R denotes the space of finite real numbers in the usual topology.

Definition 2.1. A real valued f on E is lower semi-continuous (l.s.c.) at $x = x_0$ if to every $\varepsilon > 0$ there corresponds a neighborhood $N(\varepsilon)$ of x_0 such that

(2.1) $$f(x) \geqq f(x_0) - \varepsilon$$

for $x \in N(\varepsilon)$, or, equivalently, if

(2.2) $$\underline{\lim}_{x \rightarrow x_0} f(x) \geqq f(x_0).$$

f is l.s.c. in E if it is l.s.c. in every point of E.

We recall that $\underline{\lim}_{x \rightarrow x_0} f(x) = \liminf_{x \rightarrow x_0} f(x)$ is the number λ uniquely determined by the properties:

α) to each $\eta > 0$ there corresponds a neighborhood $U(\eta)$ of x_0 such that

$$f(x) > \lambda - \eta \text{ for } x \in N(\eta) - x_0.$$

β) to each neighborhood U_1 of x_0 and to positive η we have

$$f(x_1) < \lambda + \eta$$

for some $x_1 \in U_1 - x_0$.

(1) For topological notions not defined in this paper the reader is referred to [4;I]. (Numbers in brackets refer to the bibliography at the end of this paper).

A direct consequence of this definition of <u>lim</u> is the following lemma which we will need later on

<u>Lemma 2.1.</u> Let

$$L_a = \{ x \mid f(x) \leqq a \} .$$

Then the real number b satisfies $b \geqq \underline{\lim}_{x \to x_0} f(x)$ if and only if $x_0 \subset \bigcap_{a>b} \overline{L}_a$ where as usual "$\bigcap$" denotes intersection and the bar denotes closure[2].

<u>Lemma 2.2.</u> Let f be l.s.c. on E . Then the set

$$A_r = \left\{ x \in E \mid f(x) \leqq r \right\}$$

is closed.

Proof. We prove that $A' = E - A_r$ is open . For $x_0 \in A'$ the number $\varepsilon = (f(x_0) - r) / 2$ is positive. Consequently there exists a neighborhood $N(\varepsilon)$ of x_0 for whose elements the inequality (2. 1) is true . But $f(x_0) - \varepsilon = r + \varepsilon > r$. Thus $f(x) > r$, i.e. $x \in A'$ for all $x \in N(\varepsilon)$, and A' is open .

We recall that E is called compact if every open covering of E contains a finite subcovering, and that a subset A of E is compact if A as a topological space with the relative topology generated by that of E [4;I.4.12] is compact.

<u>Theorem 2.1</u> . If E is compact and f l.s.c., then f reaches an (absolute) minimum in E.

Proof.[3] We first prove that f is bounded from below

(2) See footnote to theorem 4.4.

(3) [8 ; lemma 2.2]. The assumption made there that E is a T_1 - space is not necessary.

assume this is not true. Then none of the sets $F_n = \{x \in E \mid f(x) \leq -n\}$, $n = 1, 2, \ldots$ would be empty. Moreover $F_{n+1} \subset F_n$, and each F_n is closed by lemma 2.2. By a well known property of compact spaces [4; I.5.6], $F = \bigcap_1^\infty F_n$ would then be not empty. We thus arrived at a contradiction since $f(x) = -\infty$ for $x \in F$.

Thus $f(x)$ has a finite greatest lower bound μ. If then $G_n = \{x \in E \mid f(x) \leq \mu + \frac{1}{n}\}$, an argument analogous to the one above shows that $G = \bigcap_1^\infty G_n$ is not empty. Obviously $f(x) = \mu$ for $x \in G$. This proves the theorem.

Corollary 1. Let E be a topological space, let A be a compact subset of E, and let f be l.s.c. on A. Then f reaches a minimum on A. If f is continuous on A then it reaches also a maximum.

The last sentence follows from the remark that the continuity of f implies the lower semi continuity of $-f$.

The remainder of this section deals with convex functions. We begin by recalling the basic definitions : let E be a real linear topological Hausdorff space [4; II.1.1].

Definition 2.2. The subset A of E is called convex if for any couple x_1, x_2 of points in A the segment between x_1, x_2, i.e. the set of points $x(t) = (1-t)x_1 + t x_2$, $0 \leq t \leq 1$, is in A. With these notations a real valued function f defined on A is called convex if $f(x(t)) \leq (1-t) f(x_1) + f(x_2)$.

Theorem 2.2. Let f be defined and convex on the non empty convex open subset A of the linear topologican Hausdorff space E. We assume moreover that f is bounded on some open non empty subset of A. Then f is continuous on A.

This is part of a well known theorem [2 ; chapter 5, proposition 2].

Theorem 2.3. With the assumptions of theorem 2.2 let τ denote the topology of E, and let τ_1 be another topology of E. Suppose that for any convex set $C \subset E$.

$$(\bar{C})_{\tau} = (\bar{C})_{\tau_1} \tag{3}$$

where the bar denotes closure and where the index indicates in which topology the closure is taken. Then f is l.s.c. in the τ_1-topology [4)].

Proof. By theorem 2.2 f is continuous, and therefore l.s.c. in the τ- topology of E. We claim this implies that f is l.s.c. also in the τ_1-topology. Indeed, it is easily verified that on account of the convexity of f the set L_a of lemma 2.1. is convex. Therefore (3) is true with $C = L_a$. But this equality implies by lemma 2.1 that $(\underline{\lim}_{x \rightarrow x_0} f(x))_{\tau} = (\underline{\lim}_{x \rightarrow x_0} f(x))_{\tau_1}$. Our assertion follows now from (2) in definition 2.1.

3. Weak topology in Banach spaces. A linear topological space E is called normed if its topology is induced by a norm, i.e. by a function $\|x\|$ on E which is positive for $x \neq \theta$ the zero element, which satisfies the triangular inequality, and

4) See footnote to theorem 4.4.

the relation $||\lambda x|| = |\lambda| \, ||x||$ for real λ. A Banach space is a complete linear normed space.

A linear functional ℓ on the Banach space E is a linear map $E \to R$. The set of all continuous linear functionals becomes a Banach space E^*, the conjugate to E, if the linear operations are defined in the obvious way, and if the norm $||\ell||$ of ℓ is defined as the smallest number M satisfying $|\ell(x)| \leq M ||x||$.

<u>Definition 3.1.</u> a) The E^*-topology of E is defined as follows in terms of neighborhoods : let $\mathcal{F}_1(\theta)$ be the family of all sets W of the form

$$(3.1) \quad W = W(\ell_1 \dots \ell_n, \varepsilon) = \{x \in E \mid |\ell_i(x)| < \varepsilon, \ i = 1, \dots n\}$$

where ε is a positive number and where $\ell_i \in E^*$. Let $\mathcal{F}(\theta)$ be the family of all sets of E which contain a finite intersection of elements of $\mathcal{F}_1(\theta)$. Finally let $\mathcal{F}(x)$ be the family of all sets of the form $x + F(\theta)$ where $F(\theta) \in \mathcal{F}(\theta)$. It can be shown [4 ; V.3.3] that the elements of $\mathcal{F}(x)$ as neighborhoods of x induce the topology of a linear topological Hausdorff space which in addition is locally convex, i.e. contains a neighborhood base consisting of convex sets. This topology is the E^* topology of E. In this paper by "weak topology of E" is always meant the E^*topology of E. A set which is open in the E^*- topology is called E^*-open or weakly open. The corresponding terminology is used for all other topological notions. In contradistinction the original norm induced topology of E is often referred to as the strong topology of E

b) the E-topology of E^* is obtained from a) by interchanging the role of E and E^* and by replacing the sets (3.1) by

$$W^* = W^*(x_1, \dots x_n, \varepsilon) = \{ \ell \in E^* \,|\, |\ell(x_i)| < \varepsilon, i = 1, \dots n \}.$$

We recall the definition of a generalized sequence [4 ; I. 7. 1].

<u>Definition 3.2.</u> A partially ordered set $A = \{A, >\}$ is called directed if to every couple α, β of elements in A there exists a γ such that $\gamma > \alpha$, $\gamma > \beta$. A map $\alpha \rightarrow x_\alpha$ of A into a space E is called a generalized sequence in E. If A is the set of natural numbers in their natural order then the generalized sequence is called an ordinary sequence or simply a sequence. If E is a topological space the generalized sequence x_α is said to converge to x_0, ($x_\alpha \rightarrow x_0$), if and only if to every neighborhood N of x_0 there corresponds an α_0 such that $x_\alpha \in N$ for $\alpha > \alpha_0$. If E is a Banach space we write $x_\alpha \xrightarrow{w} x_0$ to indicate weak convergence.

<u>Lemma 3.1.</u> Let E be a Banach space and x_α a generalized sequence in E. Then

$$x_\alpha \xrightarrow{w} x_0 \tag{3.2}$$

if and only if

$$\ell(x_\alpha) \rightarrow \ell(x_0) \text{ for every } \ell \in E^*. \tag{3.3}$$

We omit the proof which follows directly from the definitions.

The following lemma expresses a well known fact [4 ; I.7.4] .

Lemma 3.2 Let E be a topological space and f a real valued function. Then f is continuous at x_0 if and only if

$$(3.4) \qquad \lim f(x_\alpha) = f(x_0)$$

for any generalized sequence in E for which

$$(3.5) \qquad x_\alpha \to x_0 .$$

This lemma applied to a Banach space E in its weak topology implies, together with lemma 3.1, immediately:

Lemma 3.3. Let E be a Banach space. Then f is weakly continuous at x_0 if and only if (3.4) is true for any generalized sequence satisfying (3.3)

Definition 3.3. The real valued f defined on a subset of the Banach space E is called weakly sequentially continuous at x_0 if (3.4) is true for any ordinary sequence for which (3.3) is true.

Since every ordinary sequence is a generalized sequence the following lemma is obvious.

Lemma 3.4. If f is weakly continuous at x_0 then it is also weakly sequentially continuous.

Theorems 3.1 and 3.2 below go in the opposite direction. We recall that a topological space is called separable if it contains a countable dense set.

Theorem 3.1. For bounded sets A in a Banach space E whose conjugate space E^* is separable the notions of weak and sequentially weak continuity coincide.

Proof. Let f be weakly sequentially continuous at x_0 and suppose it were at x_0 not weakly continuous in A. Then there would exist a positive ε such that for every weak neighborhood N of x_0 the set $N \cap A$ would contain a point x_N for which

$$|f(x_N) - f(x_0)| \geq \varepsilon \quad . \tag{3.6}$$

Now $\{x_N\}$ may be considered as a generalized sequence since the neighborhoods of x_0, ordered by inclusion, form a partially ordered set. Obviously

$$x_N \xrightarrow{w} x_0 \tag{3.7}$$

in the sense of definition 3.2. Now let R be such that $A \subset B_R = \{x \mid \|x\| \leq R\}$. The separabilitty of E^* implies that the weak topology of B_R is metric [4 ; V.5.2] . On account of (3.7) this implies that $\{x_N\}$ contains an ordinary subsequence $\{x_i\} = \{x_{N_i}\}$ $i = 1, 2, \ldots$ which converges weakly to x_0. On account of (3.6) this is a contradiction to the assumption that f is weakly sequentially continuous.

Before stating the next theorem we recall the definition of a re exive Banach space : for $x \in E$ define a function $\hat{x}$ on E^* by setting $\hat{x}(\ell) = \ell(x)$. $\hat{x}$ is linear and bounded, and therefore an element of $(E^*)^* = E^{**}$. Thus a "natural" map $k : E \to E^{**}$ is defined by $k(x) = \hat{x}$. This map is linear and isometric [4 ; II.3.19].

<u>Definition 3.4.</u> The Banach space E is called reflexive (or

regular) if the map k defined above is onto.

Theorem 3.2. For a bounded set A of a separable and reflexive Banach space E the notions of weak and sequentially weak continuity coincide.

The proof is based on the following lemma:

Lemma 3.5.[5)] If the conjugate space F^* of a Banach space F is separable then F is separable.

This is well known [4 ; II. 3. 16].

Proof of theorem 3.2. By theorem 3.1 it will be sufficient to prove that E^* is separable, and for this it is, by lemma 3.5 with $F = E^*$, sufficient to show that E^{**} is separable. But this follows from the reflexivity and separability of E.

In general the notions of weak and weak sequential continuity do not agree. This statement is true even for separable and reflexive spaces if the boundedness condition of theorem 3.2 is not satisfied. We give the following example.

Example 1. Let E be the real Hilbert space of points $x = (x_1, x_2, \ldots)$, $\sum x_1^2 < \infty$ with the usual definitions of the linear operations and of the scalar product. E is certainly separable and reflexive. Now let A be a subset of E with these properties: α) $\theta \notin A$, β) $\theta \in \bar{A}$ where the bar denotes weak closure, γ) there exists no ordinary sequence in A converging weakly to θ. (An example for such a set was given by V. Neumann [4 ; V. 7. 38].) . Now let

---

5) The idea to use this lemma for the proof of theorem 3.2 was suggested to me by A.L. Shields.

$$(3.8) \qquad f(x) = \begin{cases} 1 & \text{for } x \in A \\ 0 & \text{for } x \notin A \end{cases}$$

Then $f(x)$ is weakly sequentially continuous at θ ; for if x_n $(n = 1, 2, \ldots)$ is a sequence converging to θ then by $\gamma)$, $x_n \notin A$ except possibly for a finite number of n, and we have from (3.8) and $\alpha)$ $\lim_{n\to\infty} f(x) = 0 = f(\theta)$. On the other hand it follows from $\beta)$ that there exists a generalized sequence $\{x_\alpha\}$ in A converging weakly to θ [4; I.7.2]. For such a generalized sequence $f(x_\alpha) = 1$ by (3.8), and therefore $\lim_{x\to 0} f(x_\alpha) = 1 \neq 0 = f(\theta)$ which by lemma 3.2 shows that f is not weakly continuous at θ.

The next example does not refer to the weak topology but to the so called Γ-topology of E whose definition is obtained from the definition 3.1 a) of the weak or E^*-topology by replacing E^* by a subset Γ of E^* which is linear and total, i.e. has the property that $\ell(x) = 0$ for all $\ell \in \Gamma$ implies $x = \theta$. [4 ; V.3.2]. Correspondingly the definitions of "Γ-continuous" and "Γ-sequentially continuous" are obtained from those for weak and sequentially weak continuity resp. by replacing E^* by Γ.

Example 2. Let E be the Banach space of all bounded functions $x = x(t)$ which are defined and bounded in $0 \leq t \leq 1$, with the norm $\|x\| = \sup_{0 \leq t \leq 1} |x(t)|$. Let Γ be the set of linear combinations of elements ℓ_1 of E^* of the form $\ell_1(x) = x(t_1)$. Obviously Γ is total. Let now $B \subset E$ be the set of those $x(t)$ which belong to one of the Baire classes (see e.g; [13]), and let $B' = E - B$. Neither set is empty. We set

(3.9) $$f(x) = \begin{cases} 1 & \text{for } x \in B \\ 0 & \text{for } x \in B' \end{cases}$$

Let $x_0 \in B'$. We will show: α) f is not Γ-continuous at x_0, β) f is Γ-sequentially continuous at x_0.

α) since by (3.9) $f(x_0) = 0$, and $f(x) = 1$ for x continuous it will be sufficient to prove that every Γ-neighborhood W of x_0 contains a continuous function. We may assume that W is of the form

$$W = \{ x \in E \mid |x(t_i) - x_0(t_i)| < \varepsilon, \ 0 \leqq t_i \leqq 1; \\ i = 1, 2, \ldots n \}.$$

Then any continuous $x(t)$ which satisfies $x(t_i) = x_o(t_i)$ is contained in W.

β) let x_n be an ordinary sequence which converges to x_o in the Γ-topology, i.e. for which

(3.10) $$\lim_{n\to\infty} \ell(x_n) = \ell(x_o)$$

for every $\ell \in \Gamma$. By definition of Γ, (3.10) implies that

(3.11) $$\lim_{n\to\infty} x_n(t) = x_0(t), \quad 0 \leqq t \leqq 1.$$

We claim first,

(3.12) $$x_n \in B'$$

except possibily for a finite number of n. For otherwise there would exist an infinite subsequence $\{n_i\}$ of n such that $x_{n_i} \in B$. But this together with (3.11) for $n = n_i$ would by definition of the Baire classes imply that $x_0 \in B$ in contradiction to the choice of x_0. Thus (3.12) is proved, and we see from (3.9) that $\lim_{n\to 0} f(x_n) = 0 = f(x_0)$.

4. On the existence of extrema. The following theorem is basic

Theorem 4.1 . A Banach space E is reflexive if and only if every bounded and weakly closed subset of E is weakly compact [4 ; V.4.8] .

Also of importance is

Theorem 4.2. For convex sets in a Banach space closure and weak closure are identical [4 ; V.3.13] or [2 ; chapter IV , § 3, cor. 2] .

Corolllary to theorems 4.1 and 4.2. In a reflexive Banach space a bounded closed convex set is weakly compact.

Combining these theorems with theorem 2.1 and its corollary we obtain

Theorem 4.3. Let f be a real valued function which is defined and weakly l.s.c. on a weakly closed bounded subset A of a reflexive Banach space. Then f reaches a minimum in some point of A ; it reaches there also a maximum if it is weakly continuous. The same is true if A is bounded , closed and convex.

Combining theorem 4.2 with theorem 2.2 we obtain

Theorem 4.4 [6)] Let A be a convex closed subset of the Banach space E. Let f be defined and convex on A . Assume that A contains an interior point x_1 and a neighborhood N of x_1 such that f is bounded from above on N. Then f is weakly

6) The proof given here (as well as lemma 2.1 and theorem 2.2) is due to R. T. Rockafellar. It is contained in a letter of his to G. Minty of April 15, 1965

l.s.c. on A.

Remark. Sufficient for the lower semi-continuity of f at the point x_0 is the existence of an $\ell \in E^*$ such that $f(x) - f(x_0) \geq \ell(x - x_0)$, [8 ; theorem 4.1].

This remark may be used to obtain an alternate proof for theorem 4.4. [10; p. 154].

Combining theorem 4.4 with theorem 4.3 we obtain

Theorem 4.5. Let E, A, f satisfy the assumptions of theorem 4.4. Moreover Let A be bounded. Then f reaches a minimum on A.

5. On the relation between f and its Gâteaux differential δf.

It is clear from theorem 4.3 that it will be important to have a sufficient condition for f to be weakly continuous. The discussion of the connection between f and δf will (among other things) furnish such a condition.

First we need some preparations.

Definition 5.1. Let $x_0 \in E$, a Banach space. Let the real valued f be defined on a convex neighborhood $N(x_0)$ of x_0. Let $h \in E$ be such that $x_0 + h \in N(x_0)$, and, for $0 \leq t \leq 1$, let $\varphi(t) = f(x_0 + th)$. If the derivative $\varphi'(t)$ exists at $t = 0$, then $\varphi'(0)$ is called the Gâteaux derivative (or variation) of f at x_0 in direction h and is denoted by $\delta f(x_0; h)$

δf satisfies the relation $\delta f(x_0; \alpha h) = \alpha\, \delta f(x_0; h)$ for α real, but it is not necessarily linear in h.

Lemma 5.1 (Mean value theorem) [14; lemma 3.1]. Suppose

$\delta f(x; h)$ exists for all x, h such that x and $x+h$ are in a convex set N. Then for some t in $(0, 1)$

$$f(x+h) - f(x) = \delta f(x + th; h).$$

Definition 5.2. Let E and F be Banach spaces. Let A be a subset of E and let G be a map of E into F. Then α) G is called compact if the image under G of every bounded subset of A has a compact closure,

β) G is called completely continuous if it is compact and continuous.

Next we define the Leray-Schauder approximation for compact sets and compact mappings first used by these authors in [7].

Definition 5.3. Let F be a Banach space and B a subset of F with compact closure $\overline{B}$. Let η be a positive number and let $\ell_1, \ell_2, \ldots \ell_r$ be points of B such that for every $y \in B$

$$(5.1) \qquad \| y - \ell_i \| < \eta \text{ for at least one } i,$$

the existence of such ℓ_i being assured by the compactness of $\overline{B}$ (see e.g. [4 ; I. 6. 15]). For $j = 1, 2, \ldots r$, let μ_j be the continuous function defined by

$$\mu_j(y) = \begin{cases} \eta - \| y - \ell_i \| & \text{if } \| y - \ell_j \| < \eta \\ 0 & \text{if } \| y - \ell_j \| \geq \eta \end{cases}$$

Then

$$(5.2) \qquad a(y) = \frac{\sum_1^r \mu_j(y)\, \ell_j}{\sum_1^r \mu_j(y)} \qquad y \in B$$

is called the Leray-Schauder η approximation of the set B. Moreover if B = G(A) where G and A are as in definition 5.2 α) with A bounded, then

$$(5.3) \qquad G_\eta(x) = a(G(x)) \qquad x \in A$$

is called the Leray-Schauder approximation of the map G.

Lemma 5.2. a(y) is defined and continuous for all $y \in B$; $G_\eta(x)$ is defined for all $x \in A$, and continuous if G is completely continuous. Moreover

$$(5.4) \qquad \| y - a(y) \| < \eta, \qquad y \in B$$

$$(5.5) \qquad \| G(x) - G_\eta(x) \| < \eta, \quad x \in A .$$

All these statements follow directly from the definitions involved.

Theorem 5.1. Let the real valued f be defined in the bounded convex subset A of the Banach space E. We make the following two assumptions a) and b): a) at each $x \in A$ and for each h such that $x + h \in A$, the Gâteaux differential

$$\delta f(x; h) = G(x; h) = G(x) h$$

exists and is linear and bounded in h such that G(x) is a map $A \longrightarrow E^*$. b) this map is compact.

Then there corresponds to every positive ε a positive δ and elements $\ell_1, \ldots \ell_r$ of E^* such that

(5.7) $$|f(x+h) - f(x)| < \varepsilon \|h\| \qquad x, x+h \in A$$

for all h satisfying

(5.8) $$|l_i(h)| < \delta \|h\| , \qquad i = 1, \ldots r.$$

Proof.[7] We apply definition 5.3 and lemma 5.2 with $F = E^*$ and $\eta = \varepsilon/2$. Since

$$|G(x)h| \leq |(G(x) - G_\eta(x))h| + |G_\eta(x)h|$$

we see from (5.5) that

(5.9) $$|G(x)h| < |G_\eta(x)h| + \eta\|h\| .$$

On the other hand, from (5.3), (5.2):

$$|G_\eta(x)h| = \frac{\left|\sum_1^r \mu_j(G(x))\ell_j(h)\right|}{\sum_1^r \mu_j(G(x))}$$

$$\leqq \max_{j=1,\ldots r} |\ell_j(h)|$$

Thus from (5.9)

[7] The proof is the same as in [8 ; th. 3.3] where however the assumption is slightly stronger.

(5.10) $$\left|G(x)h\right| \leq \frac{\varepsilon \|h\|}{2} + \operatorname*{Max}_{j=1,\dots r} |\ell_j(h)|$$

and for h satisfying (5.8) with $\delta = \varepsilon/2$

(5.11) $$|G(x)h| < \varepsilon \|h\|$$

for all $x \in A$. But by lemma 5.1 this inequality implies 5.7.

Corollary to theorem 5.1. Under the assumptions of theorem 5.1 f is weakly continuous.

Indeed we have to prove : to every positive ε_1 there exists a positive δ, and elements $\ell_1, \dots \ell_r$ of E^* such that

(5.12) $$|\ell_j(h)| < \delta \qquad j = 1, \dots r$$

implies

$$|f(x+h) - f(x)| < \varepsilon_1 .$$

But this follows from (5.10) with $\varepsilon = \varepsilon_1/_{2R}$ where $R \geq \|x\|$ for $x \in A$.

This finishes the proof of Theorem 5.1 and its corollary.

Theorems in the opposite direction for general Banach spaces were proved in [1] and [5 ; th. 2] after several authors had proved such theorems for Banach spaces of a special type. See e.g. [8 ; th. 3.3] and [14 ; th. 7.2].

The following theorem and its proof are taken from [1] 8)

8) With a slight change due to the fact that in [1] the existence of a Fréchet differential (instead of only a linear Gâteaux differential) is assumed.

Theorem 5.2. Let f satisfy assumption a) of theorem 5.1. In addition we assume that

(5.13) $\|G(x)\| < M \qquad x \in A.$

Finally we assume that to $\varepsilon > 0$ there exist r elements ℓ_j of E^* such that (5.7) is true for all h satisfying

(5.14) $\ell_j(h) = 0 \qquad \|h\| < \varepsilon.$

Then G(x) is compact.

The proof is based on

Lemma 5.3. Let N be a closed linear subspace of E with a finite co-dimension. Let B_1 be the open unit ball of E. Then there exists a compact subset C of B_1 such that every $h \in B_1$ has a representation

(5.15) $h = c + n, \quad c \in C, \quad n \in N.$

For the proof we refer to [1].

Proof of theorem 5.2. If G(x) were not compact there would exist a positive number η and a sequence $x_1, x_2, \ldots$ of points in A such that

(5.16) $\|G(x_i) - G(x_j)\| > \eta, \quad i, j = 1, 2, \ldots.$

Let $\ell_1 \ldots \ell_r$ be such that (5.14) implies (5.7) with $\varepsilon = \eta/8$, and let N be the intersection of the null spaces of the ℓ_j. For $\|h\| < 1$ we then have the decomposition (5.15) which implies

(5.17) $\|n\| \leq \|c\| + \|h\| \leq 2.$

Now it is easily seen from (5.13) that the sequence of functions $\varphi_j(h) = G(x_j)h$ is uniformly bounded and equicontinuous for $h \in B_1$. Consequently there exists a subsequence $\varphi_{j_i}(h)$ which converges uniformly on the compact set C, and there exists an i_0 such that

(5.18) $$|\varphi_{j_i}(c) - \varphi_{j_k}(c)| = |G(x_{j_i})c - G(x_{j_k})c| < \frac{\eta}{2}$$

for i, $k > i_0$ and all $c \in C$. On the other hand (5.7) holds with $\varepsilon = \eta/8$ for all $n \in N$. Obviously we may in this inequality replace n by tn for $0 < t \leq 1$ to obtain

$$\frac{|f(x + t\eta) - f(x)|}{t} < ||n|| \; \eta/8$$

Letting $t \to 0$ we see that

$$|G(x)n| = |\delta f(x; n)| \leq ||n|| \; \eta/8 \text{ for all } n \in N,$$

and therefore, using also (5.17),

$$|(G(x_{j_i}) - G(x_{j_k}))n| < 2 \; ||n|| \; \eta/8 \leq \frac{\eta}{2}.$$

From this inequality together with (5.18) and the decomposition (5.15) we see that

$$|G(x_{j_i}) - G(x_{j_k})h| \leq \eta \qquad \text{for all } h \in B_1.$$

But this implies the inequality

$$|| G(x_{j_i}) - G(x_{j_k}) || \leq \eta$$

which contradicts (5.16).

6. Applications to Sobolev spaces. We start by recalling some of the basic definitions and properties of these spaces [9)]. Let Ω be a bounded open domain in the real Euclidean n-space of points $t = t_1, t_2, \dots t_n$.

We will always suppose that Ω is a Sobolev domain. This means $\Omega = \bigcup_{j=1}^{n} \Omega^j$ where each Ω^j is starlike, i.e. contains a ball B^j such that Ω^j is a star domain with respect to each point of B^j, and where the intersection $\Omega^\ell \cap (\bigcup_1^{\ell-1} \Omega_j)$ is not empty for $\ell = 2, 3, \dots n$. By "derivative" of a function $x(t)$ defined on Ω or its closure $\overline{\Omega}$ we will always mean generalized derivative as defined, e.g. in [11; p. 33]. $L_p = L_p(\Omega)$ spaces and the "p-norm" $\|x\|_p$ of an element $x \in L_p$ are defined as usual. We always assume $p > 1$.

For any positive integer ℓ the Sobolev space $W_p^\ell = W_p^\ell(\Omega)$ is then defined as the linear space of all $x \in L_p(\Omega)$ whose derivatives of order ℓ exist and are elements of $L_p(\Omega)$ with the norm given by

$$(6.1) \qquad \|x\|^p_{W_p^\ell} = \|x\|_p^p + \|x\|^p_{\ell,p}$$

where $\|x\|_{\ell,p}$ is defined as follows : let $x^{(\ell)} = (x_1^\ell, \dots x^\ell_{\sigma_\ell})$ be the vector whose components are the ℓ-th derivatives of

9) A short survey of the properties needed here is contained in [10; section 2]. For a thorough study of Sobolev spaces we refer the the reader to [12] or [11; chapter IV].

x in an arbitrary but fixed order, and denote by $\|x^{\ell}\|_E$ the Euclidean norm of this vector. Then $\|x\|_{\ell,p}$ is the p-norm of $\|x^{(\ell)}\|_E$

With these definitions the linear normed space W_p^{ℓ} can be proved to complete (see e.g. [11 ; section 112]). It is therefore a Banach space. Moreover the existence of the derivatives of order ℓ implies the existence of those of order $m < \ell$ ([11 ; section 114]) . $x^{(m)}$, σ_m are defined analogously to $x^{(\ell)}$, σ_ℓ while $x^{(0)} = x$. For later use we note that there exists a constant C such that

$$(6.2) \qquad \|x_i^{(m)}\|_p \leqq C \|x\|_{W_p^{\ell}} \qquad \begin{aligned} m &= 0, 1, \ldots \ell \\ i &= 1, 2, \ldots \sigma_m \end{aligned}$$

(see e.g. [11 ; p. 362, formula (193)] .)

As subspace of the separable space L_p the space W_p^{ℓ} is separable. Moreover it is reflexive. (see [10; section 2] and the literature quoted there.)

From these properties of W_p^{ℓ} together with the abstract results of the previous sections we are allowed to assert the facts stated in

<u>Theorem 6.1.</u> Let $A \subset W_p^{\ell}$ be bounded and weakly closed (the latter condition being satisfied if A is convex and strongly closed) . Then if the real valued f is weakly l.s.c., it reaches a minimum in A. If f is weakly continuous it reaches also a maximum in A. Moreover for any bounded A the notions of weak and sequentially weak continuity are identical (cf. th. 3.2.) .

We now derive additional results. They concern weak continuity of certain real function and compactness of certain gradients. They are all based on theorems 6.2 and 6.3 below. The proofs of these theorems are given [10 ; theorems 4.2 and 4.3].

Theorem 6.2. Let Ω be a Sobolev domain and η a positive number. Then there exists a positive constant $M = M(\Omega)$, a positive $\delta < 1$ (which may be chosen arbitrarily small) and bounded linear functionals $\lambda^j_{\beta_1 \cdots \beta_n}$ on W^l_p such that for $x \in B_{2R}$, the ball in W^l_p of center θ and radius $2R$,

$$
\begin{aligned}
(6.3)\quad & \int_\Omega \left| \frac{\partial^m x}{\partial t_1^{m_1} \cdots \partial t_n^{m_n}} \right|^p dt \\
& \leq M^p \delta^{(l-m)p} \int_\Omega \Big[\sum_{\Sigma\beta_i = l-m} \Big(\frac{\partial^l x}{\partial t_1^{m_1+\beta_1} \cdots \partial t_n^{m_n+\beta_n}} \Big)^2 \Big]^{p/2} dt \\
& + \delta^{-mp} \sum_{j=1}^{N} \sum_{k=0}^{l-m-1} \Big[\sum_{\Sigma\beta_i = k} \big(\lambda^j_{\beta_1 \cdots \beta_n}(x) \big)^2 \Big]^{p/2} + \eta, \quad \sum_{i=1}^{n} m_i = m
\end{aligned}
$$

Definition 6.1. For a positive $\delta < 1$ we denote by I_δ a closed cube of side length δ. For a Sobolev domain Ω we write $\Omega \doteq \Omega_N$ if it can be represented as the union of N cubes I^j_δ, $j=1, \ldots N$, with disjoint interiors .

We note that if $\Omega = \Omega_N$, the δ occuring in definition 6.1 may be chosen arbitrarily small, but that N depends on δ.

Theorem 6.3 . Let $\Omega = \Omega_N$. Then there exists a positive constant $M = M(\Omega)$ and bounded linear functionals $\lambda^j_{\beta_1 \dots \beta_n}$ on Ω such that (6.3) holds with $\eta = 0$.

We now draw various consequences of theorems 6.2 and 6.3.

Theorem 6.4. Let $A \subset B_r \subset W^{\ell}_p$ and let A be convex. Then for $m \leq \ell - 1$

a) for any Sobolev domain Ω, $\| x_i^{(m)} \|_p$ is weakly continuous in A.

b) if $\Omega = \Omega_N$ then to every $\varepsilon > 0$ there exists a positive $\mathcal{J}$ and elements $\ell_1, \dots \ell_r$ of $(W^{\ell}_p)^*$ such that

$$(6.4) \qquad \Big| \, \| (x+h)_j^{(m)} \|_p - \| x_j^m \|_p \Big| < \varepsilon \| h \|_{W^{\ell}_p}, \quad \begin{array}{l} x, x+h \in A \\ j = 1, 2, \dots \sigma_m \end{array}$$

for those h which satisfy

$$(6.5) \qquad |\ell_i(h)| < \mathcal{J} \| h \|_{W^{\ell}_p} \qquad i = 1, 2, \dots r$$

Proof. For a y and k in L_p

$$(6.6) \qquad \Big| \, \| y+k \|_p - \| y \|_p^p \Big| \leq \| k \|_p$$

as is seen from Minkowski's inequality. It follows that for the proof of b) it will be sufficient to prove b'); to $\varepsilon > 0$ there exist $\mathcal{J} > 0$, and $\ell_1, \dots \ell_r \in (W^{\ell}_p)^*$ such that (6.5) implies

$$(6.7) \qquad \| h_j^{(m)} \|_p < \varepsilon \| h \|_{W^{\ell}_p}.$$

Similarly for the proof of a)it will be sufficient to prove a') : to every $\varepsilon > 0$ there exist $\mathcal{J} > 0$ and $\ell_1, \dots \ell_r \in (W_p^{\ell})^*$ such that

(6.8) $$\| h_j^m \|_p < \varepsilon$$

for those h which satisfy

(6.9) $$| \ell_i(h) | < \varepsilon \qquad i = 1, \dots r \quad .$$

To prove a') and b') we use the inequality obtained from (6.3) by replacing x by h. Then the integral at the right is majorized by $\|h\|^p_{W_p^{\ell}}$ as is seen from the definition (6.1) of the W_p^{ℓ} norm. It follows that for any $\bar{\varepsilon} > 0$ the first term of our right member will be less than

$$\frac{1}{3} (\bar{\varepsilon} \, \|h\|_{W_p^{\ell}})^p$$

if we choose the positive

(6.10) $$\delta < \left(\frac{3^{1/p} M}{\bar{\varepsilon}} \right)^{\ell - m} .$$

We now keep δ fixed satisfying (6.10) . Then we choose as the required $\ell_1, \dots \ell_r$ the functionals $\lambda^j_{\beta_1 \dots \beta_n}$ occurring in (6.3). We then see that (6.5) implies that the second term at right member of (6.3) is not greater than $[\delta^{-m} q \mathcal{J} \|h\|_{W_p^{\ell}}]^p$, and that (6.9) implies for the same quantity the upper bound $[\delta^{-m} q \mathcal{J}]^p$. Here we denoted by q the value of the triple sum in (6.3) after replacing each $\lambda^j_{\beta_1 \dots \beta_n}$ by 1.

All these estimates hold for arbitrary $\bar{\varepsilon} > 0$, and η is an arbitrary positive number in case a') and equals zero in case b'). We now choose

$$\eta = \varepsilon^p/3 \ , \quad \bar{\varepsilon} < \frac{\varepsilon}{2R} \text{ in case a')}$$

$$\eta = 0 \ , \quad \bar{\varepsilon} = \varepsilon \quad \text{in case b')} .$$

We make this choice first, choose then δ according to (6.10), and then $\mathcal{J}$ such that $\delta^{-m} q \mathcal{J} < \frac{\bar{\varepsilon}}{3^{1/p}}$. Taking into account that $\| h \| \leq 2R$, we see that each term at the right member of (6.3) is less than $\frac{1}{3} \varepsilon \| h \|^p_{W^\ell_p}$ in case b'), and less than $\frac{1}{3}\varepsilon^p$ in case a'). This finishes the proof.

<u>Theorem 6.5.</u> In addition to the assumptions of theorem 6.4, b), let $\| x_i^{(m)} \|_p \neq 0$. Then the Gâteaux differential of x_i^m as map $A \to (W^\ell_p)^*$ is compact.

<u>Proof.</u> The theorem follows from the b-part of theorem 6.4 together with theorem 5.2 once it is proved that the boundedness condition (5.13) of the latter is satisfied. Now for any $y \neq 0$ in L_p

$$\delta(\| y \|_p ; h) = \int_\Omega \operatorname{sign} y \left(\frac{y}{\| y \|_p}\right)^{p-1} h \, dt \tag{6.11}$$

(see e.g. [11; p. 24].) But

$$\int_\Omega | y |^{p-1} | h | \, dt \leq \| h \|_p \| | y |^{p-1} \|_{p^1} \quad {}^{1/p + 1/p^1 = 1} \tag{6.12}$$

$$= \| h \|_p \| y \|_p^{p-1}$$

We apply this to $y = x_i^{(m)}$ with $x \in A$. Then $\|x\| \leq R$, and by (6.2), $\|y\| = \|x_i^{(m)}\| \leq CR$. Since $G(x)$ is defined by $G(x)h = \delta(\|x_i^{(m)}\|; h)$, the boundedness of G in A follows from (6.11), (6.12);

Theorem 6.6. Let again $A \subset B_R \subset W_p^{\ell}$. Then every sequence $\{x_\rho\}$ in A contains a subsequence x_{ρ_i} such that for $m = 0, 1, \dots \ell - 1$ the derivatives $x_{\rho_i}^{(m)}$ converge in L_p [10].

Proof. We consider first the case $m = 0$. Since W_p^{ℓ} is reflexive and separable it follows that $(W_\ell^p)^*$ is separable (c.f. the proof of theorem 3.2). Therefore the weak topology of B_R is metric [4; V.5.2], and since B_R is weakly compact it follows that every sequence $\{x_\rho\}$ in B_R contains a weakly convergent subsequence x_{ρ_i} [11]. Then

$$\text{(6.13)} \qquad x_{\rho_i} - x_{\rho_j} \xrightarrow{w} \theta .$$

We will show

$$\text{(6.14)} \qquad \| x_{\rho_i} - x_{\rho_j} \|_p \to 0$$

finishing the proof of our theorem for $m = 0$. To prove (6.14) we replace x by $x_i - x_j$ in (6.3) with $m = 0$. Observing

---

[10] For simplicity's sake we denote here by $x^{(m)}$ one of the m-th derivatives previously denoted by $x_i^{(m)}$ $(i = 1, \dots \sigma_m)$.

[11] This conclusion could also have been reached by the use of a more general theorem [4; V.6.1].

that $\| x_{\rho_i} - x_{\rho_j} \|_{W^{\ell}_p} \leq 2R$

we thus obtain

(6.15) $$\| x_{\rho_i} - x_{\rho_j} \|^p \leq (M \delta^{\ell}\, 2R)^p$$

$$+ \sum_{j=1}^{N} \sum_{k=0}^{\ell-1} \Big[\sum_{\Sigma\beta_1=k} (\lambda^j_{\beta_1 \dots \beta_n} (x_{\rho_i} - x_{\rho_j}))^2 \Big]^{p/2} + \eta.$$

Now if ε is a given number we choose positive η and δ such that

(6.16) $$\eta < \frac{\varepsilon^p}{3}, \quad \delta^{\ell} < \frac{\varepsilon}{2MR} \frac{1}{3^{1/p}},$$

and define the weak neighborhood W of θ by

(6.17) $$W = \Big\{ x \,\Big|\, |\lambda^j_{\beta_1 \dots \beta_n}(x)| < \varepsilon\, (3q)^{\frac{-1}{p}} \Big\}.$$

Because of (6.13) there exists an i_0 such that $x_i - x_j \in W$ for $i, j > i_0$. For such i, j we see form (6.15), (6.16), (6.17) that $\| x_{\rho_i} - x_{\rho_j} \|^p \leq \varepsilon^p$. This proves 6.14.

To extend the theorem to $m = 1, 2, \dots \ell - 1$ one has only to observe that $x^{(m)} \in W^{\ell - m}_p$ if $x \in W^{\ell}_p$ and that on account of (6.2) a set in $W^{\ell - m}_p$ which is bounded in the W^{ℓ}_p norm is also bounded in the $W^{\ell - m}_p$ norm.

As an application we mention the problem of the form

$$f(x) = \int_{\Omega} \varphi(t, x, x^{(1)}, \dots x^{(m)}, \dots x^{(\ell)}) dt = \text{Min}.$$

A treatment using the methods described in this paper or closely related ones has been given in [3], [3a], [5] and [10]: if one uses the notation

$$\varphi(x;y) = \varphi(t, x^{(1)}, \ldots x^{(\ell-1)}; y^{(\ell)})$$

$$f(x; y) = \int_{\Omega} \varphi(x(t); y(t))\, dt$$

such that $f(x) = f(x;x)$ then a convexity assumption on $\varphi(x; y)$ as function of y allows us to apply theorem 4.5 to $f(x; y)$ as function of y thus assuring its lower semi-continuity with respect to this variable. A proof for the weak continuity of $f(x; y)$ with respect to x may be based on theorem 6.2, and theorem 6.1 may then be used to assure the existence of a minimum for $f(x) = f(x; x)$.

We finally note: if $f(x; y)$ has a linear Gâteaux differential $G(x; y)h$ with respect to x (y fixed) given by

$$G(x; y) = \int_{\Omega} \sum_{m=0}^{\ell-1} \sum_{i=1}^{\sigma_m} f_{x_i^{(m)}} h_i^{(m)}\, dt \;, \text{ 12)}$$

then (under proper additional assumptions) an argument similar to the one used in the proof of statement b' in the proof of theorem 6.4 together with theorem 5.2 allows us to prove the compactness of $G(x; y)$ (as function of x).

12) For sufficient conditions that this assumption is true, see [14; 20, in particular 20.2].

BIBLIOGRAPHY

1. T. Ando, On gradient mappings in Banach spaces, Proc. Amer. Math. Soc. 12(1961), 297-299.

2 N. Bourbaki, Eléments de mathématique, livre V, Espaces vectoriels topologiques, Paris, Hermann and Co., 1953.

3. F.E. Browder, Variational methods for nonlinear elliptic eigenvalue problems. Bull. Am. Math. Soc. 71(1965), 176-183.

3a. F.E. Browder, Remarks on the direct methods of the calculus of variations, Archive for Rational Mechanics and Analysis, 20(1965), 251-258.

4. N. Dunford and J.T. Schwartz, Linear operators Part I, General theory, Interscience Publishers, New York, 1958.

5. G. Fichera, Semicontinuity of multiple integrals in ordinary form, Archive for Rational Mechanics and Analysis, 17(1964), 339-352.

6. J. Gil de Lamadrid, On finite dimensional approximations of mappings in Banach spaces, Proc. Am. Math. Soc. 13(1962), 163-168.

7. J. Leray and J. Schauder, Topologie et équations fonctionnelles, Annales Scientifiques de l'Ecole normale supérieure (3), 51(1934), 45-78.

8. E.H. Rothe, Gradient mappings and extrema in Banach spaces, Duke Math. J., 15(1948), 421-431.

9. ...... A note on the Banach spaces of Calkin and Morrey, Pacific J. Math., 3(1953), 493-499.

10. ......An existence theorem in the calculus of variations based on Sobolev's imbedding theorems, Archive for Rational Mechanics and Analysis 21(1966), 151-162.

11. V.I. Smirnov, A course of higher mathematics, vol. V, Pergamon Press, 1964 (translated from the Russian edition of 1960).

12. S.L. Sobolev, Applications of functional analysis in mathematical physics, Amer. Math. Soc. 1963 (translated from the Russian edition of 1950).

13. C. de la Vallée Poussin, Intégrales de Lebesgue, fonctions d'ensemble, classes de Baire, Paris, Gauthiers-Villars, 1916.

14. M.M. Vainberg, Variational methods for the study of non-linear operators, Holden-Day, 1964(translated from the Russian edition of 1956).

The University of Michigan
Ann Arbor, Michigan

CENTRO INTERNAZIONALE MATEMATICO ESTIVO
(C.I.M.E.)

E. O. ROXIN

PROBLEMS ABOUT THE SET OF ATTAINABILITY

Corso tenuto a Bressanone dal 10 al 18 giugno

1966

PROBLEMS ABOUT THE SET OF ATTAINABILITY

by

E.O. ROXIN

I. Control systems. Attainable set.

1. Control systems. In this series of lectures we will consider "control systems" described by differential equations of the type

(I.1.1) $$dx/dt = x' = f(t, x, u) ,$$

where t is a real variable representing the time, x is an n-vector determining the instantaneous state of the (physical) system, and u is an m-vector corresponding to the instantaneous action of a certain control mechanism. This control action is supposed to be adjustable as a function of time $u(t)$, or of the state $u(x)$, or of both; $u = u(t, x)$. Once the control action is prescribed, equation (I.1.1) is a differential equation governing the evolution of the system. A typical problem of control theory, is how to choose the control law $u(t, x)$, in order to achieve a certain goal (for example to reach a given point in minimum time).

In these lectures we will consider only problems in which it is assumed that the control action $u(t, x)$, even if not determined from the beginning, can be adjusted in an exact manner. In other words, we will not be concerned about "stochastic systems", where some "random functions" (which are neither exactly known nor adjustable) influence the evolution of the system.

Some particular cases of (I.1.1), given in the following, have especial importance.

In the case

(I.1.2) $$x' = f(t, x) + G(t, x) \cdot u ,$$

where G(t, x) is a nxm matrix, the equation is <u>linear in the control</u>. If it is also time-independent, we obtain

(I.1.3) $$x' = f(x) + G(x) \cdot u .$$

Equations (I.1.1) and (I.1.2) are really equivalent, one can for example write (I.1.1) in the form

(I.1.4) $$x' = f(t, x, 0) + (f(t, x, u) - f(t, x, 0)) = \\ = f_o(t, x) + v .$$

Here $f_o(t, x) = f(t, x, 0)$ gives the equation without control action, and $v = v(t, x)$ is the new control parameter ; the matrix G(t, x) in (I.1;2) is now simply the identity .

The <u>linear case</u>

(I.1.5) $$x' = A(t) x + B(t) \cdot u ,$$

where A is an nxn matrix, B an nxm matrix, is also of great importance. If A and B are constant matrices, we obtain the simplest case, which is still of sufficient practical importance and theoretical interest, to be the source of many problems.

<u>2. Admissible controls</u>. In equations (I.1.,q) to (I.1.5), if the control u is thought of as a function of the state : $u = u(x)$ or $u = u(t, x)$, serious questions arise about existence and uniqueness of solutions. We prefer, therefore , to consider u as a function of time : $u = u(t)$, which is chosen, in general, subject to certain restrictions but otherwise arbitrary. For any particular solution x(t) , there is, of course, no difference in considering $U = u(x(t)) = u_1(t)$.

E. O. Roxin

In order to guarantee the existence and uniqueness of solutions, it is generally required that the function f(t, x, u) in (I, 1, 1) be continuous in (t, x, u), lipschitzian in x. In most cases, these requirements can be relaxed, according to the classical theorem of Carathéodory, assuming that with respect to t, f(t, x, u) is only measurable and there is an integrable function of t only, m(t), majorizing f(t, x, u). In this last case, the solutions x(t) are understood in the "generalized" sense, of absolutely continuous functions which satisfy the differential equation almost everywhere. For the classical existence and uniqueness theorems of differential equations, the reader is referred, for example, to the treatise of Coddington and Levinson [1] (*)

Under these conditions, the mildest natural requirement for the control function u(t), in order to insure the existence of solutions of (I. 1. 1), is the condition that u(t) be integrable. To require that u(t) be continuous turns out to be too restrictive for many important applications (where the control function have discontinuities of the "jump" type). The piecewise continuity of u(t) seems the most natural requirement from the physical point of view.

An additional restriction is the assumption that u(t) is bounded by a given constant, or, in general, that the range of u(t) belongs to a given compact set U of the real m-dimensional space R^m. We will in general refer to this condition, saying that the control is "bounded". In many important cases, this

---

(*) References [1], [2] etc. refer to the literature given at the end.

boundedness, condition is of decisive importance, because it may turn out that in order to maximize (or minimize) certain functionals, the control action should be as large as possible. In these cases, if there is no bound for the control, no optimal control exists.

The requirements about the control functions, define the class K of "admissible" controls. It is important to specify this class in the formulation of any control problem, because the solution may depend on it.

The geometric interpretation of the condition $u(t) \in U$ for each t, is the following. At every point (t, x) in $n+1$ space, there is not one but a whole set of possible directions of dx/dt, and this set is precisely

$$C = C(t, x) = f(t, x, U) = \left\{ f(t, x, u) \; ; \; u \in U \right\} . \tag{I. 2. 1}$$

The solution curves of the control equation (I. 1. 1) have the property that at each point (t, x), the tangent line determined by dx/dt belongs to the set $C(t, x)$. More precisely: $x(t)$ should be absolutely continuous and at almost every t, the condition $x'(t) \in C(t, x(t))$ should hold.

3. Contingent and paratingent equations. Let $X \subset R^n$ be a given set, and 0 a limit point of X. In the terminology of G. Bouligand [1] the contingent of the set X at the point 0 is defined as the set of straight lines through 0, such that a line $0Q$ belongs to the contingent of X if and only if there is a sequence $P_i \in X$ $(i=1, 2, 3, \ldots)$, $P_i \neq 0$, $P_i \to 0$ for $i \to \infty$, and line $0P_i$ tends to $0Q$ for $i \to \infty$.

E. O. Roxin

Under the same conditions, the <u>paratingent</u> of X at 0 is defined as the set of straight lines $0Q$ such that there are two sequences $P_i \in X$, $Q_i \in X$, $(i=1, 2, 3, \dots)$, $P_i \neq Q_i$, $P_i \to 0$, $Q_i \to 0$ and the line $P_i Q_i \longrightarrow 0Q$ for $i \to \infty$.

If the set X is a continuous curve $x(t)$, the contingent and paratingent at the point $x(t_o)$ are defined by sequences $t_i \to t_o$, $t_i' \to t_o$, and one can define the left and right contingent and paratingent, similar to the left and right derivatives of $x(t)$.

Marchaud ([1], [2], [3]) and Zaremba ([1], [2]) considered "contingent" and "paratingent" equations, i.e. expressions of the form:

Contingent (paratingent) of $x(t)$ at $t_o \subset C(t_o, x(t_o))$,

where $C(t,x)$ is given as a function of t and x. The curve $x(t)$ is a solution of the contingent (or paratingent) equation, if it satisfies the mentioned condition at every point. The set $C(t,x)$ is considered as a set of stright lines (a "cone" or "half-cone", as Marchaud considers it).

Practically equivalent, but more analytical, is the formulation considering, instead the straight lines, their difference quotient $\Delta x / \Delta t$. In this case, $C(t,x) \subset R^n$ is the set of all possible limits of the difference quotient $(x(t_i) - x(t_o)) / (t_i - t_o)$, for $t_i \to t_o$ (and similarly for the paratingent equation). We will designate by D^* this set of all possible limits, and call it the contingent of $x(t)$ at t_o. The contingent equation is then formulated as follows:

$$D^* x \subset C(t, x) \quad \text{(I. 3. 1)}$$

Marchaud and Zaremba realized that the most important results are obtained under the following conditions:

i) For every $(t,x) \in R^{n+1}$, the set-valued function $C(t,x)$ is compact.

(ii) For every $(t,x) \in R^{n+1}$, $C(t,x)$ is convex.

(iii) $C(t,x)$ is uniformly bounded, i.e. $C(t,x) \subset K$, where K is a compact set independent of (t,x).

iv) As a function of (t,x), $C(t,x)$ is upper semicontinuous (sometimes called "upper semicontinuous by inclusion"), i.e. given $C_o = C(t_o, x_o)$, for every $\varepsilon > 0$ there is a $\delta > 0$ such that $|t - t_o| < \delta$, $|x - x_o| < \delta$, imply

$$C(t,x) \subset U_{\varepsilon}(C_o) = \left\{ x \in R^n ; \text{distance } (x, C_o) \leq \varepsilon \right\}.$$

(This hypothesis was really made by Zaremba, Marchaud considered the more restrictive case of $C(t,x)$ continuous).

In the following, a set valued function $C(t,x)$ with the mentioned properties, will be called <u>"orientor field"</u> or simply a "field" (as generalization of a "vector field").

In case the field $C(t,x)$ is not defined on the whole R^{n+1}, but only on some domain $D \subset R^{n+1}$, it is possible to generalize the results we will see later, in a straightforward manner, for the interior points of D. For simplicity, here we will refer to the case in which $C(t,x)$ is defined on the whole R^{n+1}.

Some basic results obtained by Marchaud and Zaremba under the above mentioned conditions, are the following:

a) Every solution of (I.3.1), this is every continuous $x(t)$ satisfying $D^* x(t)_{t=t_o} \subset C(t_o, x(t_o))$ for every t_o in the considered time interval I, satisfies a uniform Lipschitz condition, and therefore is absolutely continuous and rectifiable (considered as well in R^n as in R^{n+1}).

b) If a sequence $x_i(t)$ (i=1, 2, 3, ...) of solutions of (I.3.1) tends to the limit x(t), then x(t) also satisfies (I.3.1).

Note : as C(t, x) is contained in a fixed compact K, any sequence $x_i(t)$ of solutions is equicontinuous (in any finite interval). By Arzela's theorem, such sequence contains, therefore, a subsequence which converges to some limit x(t), and now we see that this limit x(t) is also a solution of the contingent equation.

c) Given an initial point (t_o, x_o), there is at least one solution x(t) of (I.3.1) satisfying $x(t_o) = x_o$.

If C(t, x) is assumed to be continuous in (t, x), and to have a non-empty interior (in R^n), then it is quite easy to prove the existence of a solution of (I.3.1) passing through a given (t_o, x_o) (it is possible to choose the solution as a poligonal path of sufficiently small straight pieces). In order to obtain the general result given above, the following auxiliary theorem has to be used.

d) Given the field C(t, x) which satisfies properties (i), (ii), (iii), (iv), there is a sequence of fields $C_i(t, x)$ (i=1, 2 3, ...), also satisfying conditions like (i), (ii), (iii), (iv), and such that $C_i(t, x)$ is continuous in (t, x), $C_i(t, x) \supset C_{i+1}(t, x)$, C(t, x) is contained in the interior of every $C_i(t, x)$ (therefore every $C_i(t, x)$ has a non-empty interior), and for $i \to \infty$, lim $C_i(t, x) = C(t, x)$.

It should also be noted that under the above conditions, the solutions of the contingent equation

$$\text{conting. } x \subset C(t, x)$$

are the same as the solution of the paratingent equation

$$\text{parating. } x \subset C(t, x) \quad .$$

Later we will see other results of Marchaud and Zaremba, concerning the "attainable set".

More recently, Ważewski ([1] to [9]) considered contingent equations and control systems, proving that under the above conditions the following statements are equivalent :

(I. 3. 2. a) $\begin{cases} x(t) \text{ is continuous in } t \in I \\ D^{*} x \subset C(t, x(t)) \text{ for every } t \in I, \end{cases}$

and

(I. 3. 2. b) $\begin{cases} x(t) \text{ is absolutely continuous in } t \in I, \\ x'(t) \in C(t, x(t)) \quad \text{a.e. } I \text{ (almost everywhere in I).} \end{cases}$

Indeed, we get (b) from (a) because any solution of a contingent equation is absolutely continuous and therefore has a unique derivative a.e. Where this derivative exists, $x'(t) = D^{*}x(t)$.

To obtain (a) from (b), consider

$$x(t) - x(t_o) = \int_{t_o}^{t} x'(s)\, ds,$$

x(t) being absolutely continuous. Given $\varepsilon > 0$, for $|t - t_o|$ sufficiently small,

$$x'(t) \in U_{\varepsilon}(C_o)$$

according to property (iv) . As $U_{\varepsilon}(C_o)$ is compact and convex, the lemma which follows assures that

$$\frac{x(t) - x(t_o)}{t - t_o} \in U_\varepsilon(C_o)$$

Passing to the limit, the desired result $D^* x \subset U_\varepsilon(C_o)$ is obtained for every $\varepsilon > 0$.

Lemma : If $y(t)$ is measurable in $[a, b]$, and $y(t) \in K$, where K is a compact and convex set, then

(I. 3. 3) $$Y = \frac{1}{b-a} \int_a^b y(t)\, dt$$

also belongs to the set K.

The proof is immediate considering the convex set K determined by its support function (Bonnesen-Fenchel, [1]) $f(p)$: a point $z \in R^n$ belongs to K if and only if for every $p \in R^n$, the scalar product

$$p.z \leq f(p)$$

Multiplying both sides of (I. 3. 3) by an arbitrary p, we obtain

$$p.Y = \frac{1}{b-a} \int_a^b p.y(t)\, dt \leq \frac{1}{b-a} \int_a^b f(p)\, dt = f(p) .$$

4. Attainable set. Given a control equation

(I. 4. 1) $$x' = f(t, x, u)$$

(x = n-vector, u = m-vector), with the restriction

(I. 4. 2) $u(t) \in K$ = class of admissible controls ;

and given the initial condition

(I.4.3) $$x(t_o) = x_o ,$$

there are, in general, a whole set of different solutions of (I.4.1) and (I.4.2) which satisfy (I.4.3). Sometimes it is useful to distinguish the solution x(t) corresponding to a certain control u(t), by writing $x_u(t)$.

We define the <u>attainable set</u> from (t_o, x_o), at $t = t_1$, as the intersection of all solutions of (I.4.1-2-3) with the hyperplane $t = t_1$, and we designate it by $F(t_1, t_o, x_o)$:

(I.4.4) $$F(t_1, t_o, x_o) = \left\{ x = x_n(t_1) ; u \in K , x_u(t_o) = x_o \right\} .$$

In the classical theory of dynamical systems, the notation $F(t, t_o, x_o)$ is used for the value at t of the (unique) solution starting at (t_o, x_o). Definition (I.4.4) is a straightforward generalization of this classical notation, because when there is no control action $(x' = f(t,x))$, both definitions coincide.

There is a certain similarity between the solution of a control equation (I.4.1-2) and the solutions of a differential equations

(I.4.5) $$x' = f(t, x)$$

in the case where f(t x) satisfies sufficient conditions for existence but not for uniqueness of solutions. Then, also, through the initial point (t_o, x_o) there are, in general, many solutions. The analogous to the attainable set is in this case the "integral funnel"

(I.4.6) $$F(t_1, t_o, x_o) = \left\{ x(t_1) ; x(t) \text{ satisfies } (I.4.5) , x(t_o) = x_o \right\} .$$

Some theorems about attainable sets of control systems, can be regarded as generalizations of theorems concerning the integral funnel for differential equations without uniqueness, as we will see later.

It is also useful to consider the attainable set in the (t, x) - space R^{n+1}:

(I. 4. 7) $$\Phi(t_o, x_o) = \left\{ (t, x_u(t)) \; ; \; t \in R \; , \; x_u(t_o) = x_o \; , \; u(t) \in K \right\} .$$

Evidently $F(t, t_o, x_o)$ is the section of $\Phi(t_o, x_o)$ with the plane t.

The difference between the functions $F(t, t_o, x_o)$ and $\Phi(t_o, x_o)$ is the difference between considering the value of a function at a particular point, and the function as a whole.

The projection of the set $\Phi(t_o, x_o)$ on the space R^n,

(I. 4. 8) $$\left\{ F(t, t_o, x_o) \; ; \; t \in R \right\} = F(R, t_o, x_o) \; ,$$

which is the set of all points attainable at some unspecified time, is sometimes of importance;.

Given a control equation (I. 4. 1-2-3), the problem of determining the attainable set, i.e. the function $F(t, t_o, x_o)$, is of fundamental importance in control theory. For the particular case of differential equations (without control), it reduces to the determination of the general solution of (I. 4. 5), From this fact, it is obvious that one cannot expect to solve this problem explicitly, in general.

E. O. Roxin

II. Properties of the attainable set.

1. Notation. We consider the control differential equation

(II. 1. 1) $x' = f(t, x, u)$ (x = n-vector, u = m-vector)

with the class of admissible controls

(II. 1. 2) $K = \left\{ u : I \to R^m ;\ u(t) \text{ measurable},\ u(t) \in U(t, x) \right\}$,

where I is a given time-interval (finite or infinite).

A solution $x_u(t)$, corresponding to the control $u(t) \subset K$, is an absolutely continuous function which satisfies (II. 1. 1) almost everywhere in I (a. e. I) .

The attainable set at t_1 from (t_o, x_o) is defined by

(II. 1. 3) $F(t_1, t_o, x_o) = \left\{ x = x_u(t_1) ;\ u, x_u \text{ satisfy (II. 1. 1-2)(a. e. I)},\ x_u(t_o) = x_o \right\}$

The solution curves x(t) of (II. 1. 1-2) are often referred to as the "trajectoires" of the equation (II. 1. 1-2) . If should be noted that, in general, a trajectory x(t) can correspond, as solution of (II. 1. 1-2) , to more that one control u(t) ; this happens when the value of u is not uniquely determined by the values of t, x and f(t, x, u) .

2. Some examples. Under very general conditions, the properties of the attainable set of a control system, can show great variations. We will see this in some examples.

Example 1. Consider, in 2 dimensions, the system

$$
\text{(II. 2. 1)} \begin{cases} x' = u \dfrac{1 + x^2}{1 + x^2 y^2} \\[2ex] y' = \dfrac{(1 + y)^2}{1 + x^2} - u \dfrac{2 x y (1 + y)}{1 + x^2 y^2} \end{cases}
$$

with the condition

$$u(t) \in [-1, +1].$$

The function

$$p = \frac{y(1 + x^2)}{1 + y}$$

satisfies, according to (II. 2. 1),

$$p' = dp/dt = 1 .$$

Therefore

$$p(t) = p(0) + t ,$$

and the attainable set at $t = t_1$ is a subset of the curve in x, y space (see figure II. 1)

$$\frac{y (1 + x^2)}{1 + y} = \frac{y_o (1 + x_o^2)}{1 + y_o} + t_1 = \text{const} .$$

Starting, for example, at $x_o = y_o = 0$, the attainable set $F(t, 0, 0)$ will be symmetric with respect to the y-axis (changing u in -u, x will change in -x and y will remain unchanged), and for $t \geq 1$, the attainable set will be symmetrically located on both branches of the corresponding curve $p(t) = t$. Therefore it will be a not - connected set .

<u>Example 2.</u> On the torus

$$(\text{II. 2. 2}) \qquad \begin{cases} x = (a + R \cos \varphi) \cos \lambda \\ y = (a + R \cos \varphi) \sin \lambda \\ z = R \sin \lambda \end{cases}$$

consider φ, λ as coordinates, and take the control system

$$(\text{II. 2. 3}) \qquad \begin{cases} \varphi' = u \\ \lambda' = k.u \end{cases}$$

with k irrational, and $|u| \leqslant 1$, The trajectories are helices which are dense on the surface of the torus. For every finite value of t, the attainable set $F(t, t_o, x_o)$ is a finite line-segment, but the attainable set for all t (i.e. $F(R, t_o, x_o)$) is dense on the surface of the torus.

This last result can be obtained even for an attainable set for a finite time, $F(t, t_o, x_o)$, considering instead of (II. 2. 3), the system

$$(\text{II. 2. 4}) \qquad \begin{cases} \varphi' = \tan u \\ \lambda' = k. \tan u, \end{cases}$$

with $|u| < \frac{\pi}{2}$.

In these examples, the control system is defined on a surface different from the R^n. Nevertheless, it is not difficult to see that in the (x, y, z)-space R^3, toroidal coordinates R, φ, λ can be taken in such a way that for R = const., a torus like (II. 2. 2) results (figure II. 2). System (II. 2. 4) should then be redefined, for example, as

$$\text{(II.2.5)} \qquad \begin{cases} \varphi' = \dfrac{1}{R} \tan u \\ \lambda' = \dfrac{k}{R} \tan u \\ R' = 0, \end{cases}$$

again with $|u| < \frac{\pi}{2}$. (The factor $1/R$ avoids the infinite speed when $R = \infty$) .

Example3. (Roxin, [3])- . Consider the system

$$\text{(II.2.6)} \qquad \begin{cases} x' = 1 \\ y' = \dfrac{|u|}{1+z^2} \\ z' = \dfrac{u}{1+z^2}, \end{cases}$$

with $|u| \leqslant 1$. The set $f(x, U)$ consists of two segments (as shown in figure II. 3) :

$$x' = 1, \quad y' = |z'| \leqslant \frac{1}{1+z^2}$$

For this system, the attainable set $F(t, t_o; x_o)$ is not closed. In order to see this, it is sufficient to consider the different trajectories which start from the origin and reach the line $y=1$, $z = 0$. The projection of these trajectories, on the y, z plane, are broken lines which make constantly an angle of $\pm 45^o$ with the y axis.

The value of x when the line $y = 1$, $z = 0$, is reached, equal to the time necessary to reach it, will be smaller for those trajectories , where z^2 is smaller, according to the second of equations (II. 2. 6) . The lower limit of this time corresponds to the "quasi-trajectory" $z = 0$, which is not an admissible trajectory, because it does not correspond to any control $u(t)$. Therefore the limit of attainable point does not belong to the attainable set, which is therefore not closed.

3. The closedness of the attainable set. For many problems in optimal control theory, the fact that the attainable set in R^{n+1} is closed, is of decisive importance. Indeed, take for example the time optimal control problem. This problem consists in giving a control equation like (II. 1. 1-2), two points x_o, x_1 in R^n, an initial time t_o ; with these data try to find the "time optimal control" u(t) and the corresponding trajectory x(t) which satisfy (II. 1. 1-2), for which $x(t_o) = x_o$ and $x(t_1) = x_1$ for minimum $t_1 > t_o$. Loosely speaking, the point x_1 has to be reached in minimum time, starting from x_o at t_o.

Geometrically, the point $(t_1, x_1) \in R^{n+1}$ is determined by the intersection of the half-line : $x=x_1$, $t \geq t_o$, with the attainable set in R^{n+1} :

$$\text{(II. 3. 1)} \qquad \Phi^+(t_o, x_o) = \left\{ (t, x) \; ; \; t \geq t_o, \; x \in F(t, t_o, x_o) \right\} .$$

If this intersection is not empty (if "it is possible to reach x_1 from x_o, starting at t_o"), and if the set $\Phi^+(t_o, x_o)$ is closed, then there is some "minimum time" t_1 (for reaching x_1), and the point $(t_1, x_1) \in R^{n+1}$ belongs to the attainable set $\Phi^+(t_o, x_o)$, i.e. there is some trajectory x(t) of (II. 1. 1-2), going from $x(t_o) = x_o$ to $x(t_1) = x_1$.

The closedness of the attainable set in R^{n+1} gives therefore an existence theorem of optimal controls and optimal trajectories.

The first results in this questions were obtained by Marchaud ([3]) and Zaremba ([2]), for contingent equations, under the hypothesis stated in chapter I, § 3 (conditions (i) to (iv)). We will here continue the list of results of Marchaud and Zaremba given there.

Under the hypothesis of chapter I, § 3, for equation (I. 3. 1):

e) The attainable set in R^{n+1}, $\Phi(t_o, x_o)$, is closed. Even more: the set

$$\Phi_{(t_1, t_2)}(t_o, x_o) = \Phi(t_o, x_o) \cap \left\{ (t, x) ; \ t_1 \leqslant t \leqslant t_2 \right\}$$

is compact.

The proof depends basically on the fact that a sequence of solutions of the contingent equation (I. 3. 1) has a convergent subsequence, and its limit is also a solution.

f) The set $\Phi_{(t_1, t_2)}(t_o, x_o)$ is an upper semicontinuous function of (t_o, x_o).

From this result, it is easily obtained that also $F(t, t_o, x_o)$ is upper semicontinuous in (t_o, x_o).

Properties (e) and (f) are really proved by Zaremba for the more general "attainable set from a given set"

$$\Phi_{(t_1, t_2)}(A) = \bigcup_{(t_o, x_o) \in A} \Phi_{(t_1, t_2)}(t_o, x_o) ,$$

where A is a compact set in R^{n+1}.

g) The set $F(t, t_o, x_o)$ is connected.

Also in this case it is true that, more in general, the set

$$F(t, A) = \bigcup_{(t_o, x_o) \in A} F(t, t_o, x_o)$$

is a continuum (closed and connected) if $A \in R^{n+1}$ is a continuum.

These theoremsare, in some sense, a generalization of more classical results about differential equations which satisfy conditions insuring the existence but not the uniqueness of solutions. For comparison, we will state briefly :

<u>Kneser's theorem</u> (Kneser, [1]) : Let $f(t,x)$ be continuous on the cylinder $t_o \leqslant t \leqslant t_1$, $|x - x_o| \leqslant a$, let $|f(t,x)| \leqslant M$, $M \leqslant a / (t_1 - t_o)$ and $t_o \leqslant \tau \leqslant t_1$. Finally, let

$$F_\tau = \left\{ x(\tau) \,;\, x(t) \text{ satisfies } x' = f(t,x) \,,\, x(t_o) = x_o \right\}$$

(attainable set at $t = \tau$). Then F_τ is a continuum (closed and connected) .

A proof of this theorem can be seen in Hartman [1] . It is also a consequence of Zaremba's result given above, for the case in which the set $C(t,x)$ in (I.3.1) reduces to a single point $f(t,x)$.

The problem of the closedness of the attainable set in R^{n+1} $\Phi(t_o, x_o)$, became important , recently, with the development of the mathematical theory of control systems and optimal control. Existence theorems of optimal control, closely related to the more abstract problem of closedness, were given for special differential systems by Bushaw ([1]) , Bellman et al. (Bellman, Glicksberg and Gross, [1]) and LaSalle ([1] , [2]) . Even if Pontryagin's Principle (see Pontryagin, Boltyanskii, Gamkrelidze, Mischchenko, [1]) is only a necessary condition and therefore gives no indication about the existence of optimal controls, practically it allows us to solve the problem of existence in many cases . The general linear case, where the equation is of the form

(II. 3. 2) $$x' = A(t) + B(t)\ u\ ,$$

with

(II. 3. 3) $$t \in R\ , \quad x \in R^n\ , \quad u(t) \in U = \text{compact set in } R^m\ ,$$

was solved by Markus and Lee ([1]) .

Filippov ([2]) gave the general existence theorem for optimal controls :

<u>Filippov's theorem</u> : Let the control differential equation be

(II. 3. 4) $$x' = f(t, x, u)\ ,$$

where x is an n-vector, u an m-vector , t the real variable/ The class of admissible controls is

(II. 3. 5) $$K = \left\{ u(t) \text{ measurable; } u(t) \in U(t, x) \subset R^m \right\} .$$

The function $f(t, x, u)$ is assumed to be continuous in (t, x, u) , continuously differentiable in x, and such that for some constant k

(II. 3. 6) $$x . f(t, x, u) \leqslant k\ (|x|^2 + 1)\ (x . y = \text{scalar product})-$$

for all t and x and all $u \in U(t, x)$. The set $U(t, x)$ is assumed to be compact, and so will be its continuous image

(II. 3. 7) $$C(t, x) = f(t, x, U(t, x)) = \left\{ f(t, x, u)\ ;\ u \in U(t, x) \right\} .$$

$U(t, x)$ is also assumed to be upper semicontinuous in (t, x) , and the same applies to its image $C(t, x)$. Finally; the crucial assumption is made, that $C(t, x)$ is convex for all (t, x) . Under these conditions, Filippov's theorem asserts that, given t_o, x_o and x_1, if there is some $u(t) \in K$ such that for the corresponding trajectory $x(t)$, $x(t_o) = x_o$, $x(t_1) = x_1$ for some $t_1 > t_o$, then there is also a "time optimal control" $u_m(t)$ such that the corresponding optimal trajectory $x_m(t)$ with $x_m(t_o) = x_o$, satisfies $x_m(t_m) = x_1$ with minimum $t_m \geq t_o$ (for any other trajectory $y(t)$ with $y(t_o) = x_o$, $y(t_2) = x_1$ for some $t_2 \geq t_o$, necessarily $t_2 \geq t_m$).

We will not go into the details of the proof.

Roxin ([3]) proved independently a property which is basically the same, even if the assumptions are slightly different. These assumptions are :

i) The set U is constant and compact :

ii) the function $f(t, x, u)$ is continuous in (x, u) for fixed t, measurable in t for fixed (x, u) ;

iii) a Lipschitz condition holds : $|f(t, x, u) - f(t, \bar{x}, u)| \leq K|x - \bar{x}|$;

iv) for all $u \in U$ (uniformly), $|f(t, x, u)| \leq \mu(t) \cdot g(|x|)$, where the function $\mu(t)$ is integrable in every finite interval, $g(|x|)$ is bounded in every bounded region of x-space, and $g(|x|) = 0(|x|)$ for $|x| \longrightarrow \infty$;

v) the set $f(t, x, U) = \{ f(t, x, u) ; u \in U \}$ is convex for each (t, x).

Under these conditions, the result is given in the more general way, that the attainable set in R^{n+1} : $\Phi(t_o, x_o)$ is closed. The existence of the time optimal solution is then an immediate consequence.

It should be noted that these theorems are practically included in Zaremba's result given above, at the beginning of this paragraph, except for one step : the equivalence between control differential equations and contingent equations. This equivalence was proved by Ważewski in 1960 ([1], [2], [3]).

<u>Ważewski's theorem</u> : Consider the control differential equation (II.1.1) with condition (II.1.2) . Consider also the contingent equation

(II.3.8) $$D^* x \subset C\ (t, x),$$

where $D^* x$ is the contingent of $x(t)$, and this $x(t)$ is a solution of (II.3.8) if it is continuous and for every t (in the considered interval) , (II.3.8) holds. This contingent equation will be called "associated" to the control equation (II.1.1-2) , if

(II.3.9) $$C(t, x) = f(t, x, U(t, x)) .$$

Assume : (hypothesis H_1) :

i) For some compact interval $I = [0, a]$ and some compact $K \subset R^m$, the function $f(t, x, u)$ is continuous on an open set containing $W = I \times B(r) \times K$, where $B(r)$ is the n-dimensional ball of center 0 and radius r :

ii) there is $M > 0$ such that $|f(t, x, u)| < M$ for $(t, x, u) \in W$;

iii) $U(t, x)$ is compact and $U(t, x) \subset K$ for $(t, x) \in I \times B(r)$;

iv) $U(t, x)$ is upper semicontinuous on $I \times B(r)$;

v) h = minimum $(a, r/M)$.

Assume also (hypothesis H_2) :

vi) The set $C(t, x) = f(t, x, U)$ is convex.

Under these conditions, the control equation (II.1.1-2) and the associated contingent equation (II.3.9) are equivalent, i.e. every solution $x(t)$ of one of these equations, is also a solution of the other.

According to what has been said in chapter I, §3, given a solution $x(t)$ of the contingent equation, it is absolutely continuous and satisfies

$$x'(t) \in C(t, x(t)) \tag{II.3.10}$$

almost everywhere. In order to prove that this $x(t)$ is a trajectory of the control system (II.1.1-2), the existence of a corresponding measurable control function $u(t)$ has to be shown. This is by no means trivial, because $x(t)$ determines (a.e.)

$$x'(t) = f(t, x, u), \tag{II.3.11}$$

but the values of t, $x(t)$ and $f(t, x(t), u)$ do not determine, in general, the value of u. Condition (II.3.10) merely proves that there is some value of u satisfying the condition (II.3.11). This value of u is, in general, not unique, and the fact that for each t it can be chosen in such a manner that the resulting $u(t)$ be measurable, is not obvious at all.

Ważewski solves this problem, taking for each value of t, as $u(t)$, the so called "lexicographic maximum" of the set

$$N(t) = \left\{ u \in U(t, x(t)) \; ; \quad f(t, x(t), u) = x'(t) \right\}. \tag{II.3.12}$$

This lexicographic maximum is obtained as follows; as the set $N(t)$ is compact, the first coordinate u_1 of $u = (u_1, u_2, \ldots, u_m)$ has some maximum in N. Let $N_1 \subset N$ be the subset of values of $u \in N$, such that their first coordinate takes this maximum value u_{1m}. Then N_1 has at most dimension $m-1$. Similarly, on N_1 the second coordinate u_2 has a maximum u_{2m}, and let N_2 be the subset of N_1 where $u_2 = u_{2m}$. Continuing this way, the set N_m consists of a single point, which is precisely the lexicographic maximum. Ważewski proved that this lexicographic maximum is a measurable function of t. We have so obtained an admissible control u(t) for the trajectory x(t). The rest of the proof of Ważewski's theorem offers no great difficulty.

4. Chattering, sliding regime and quasitrajectories. In the previous paragraph, the existence of optimal controls was proved under the assumption that the image C(t, x) of the "control domain" U(t, x) is convex. This "convexity" condition is essential, in the sense that without it, it is easy to obtain examples in which the desired result is no longer true (see § 2, example 1). For special cases (for example linear systems) the convexity condition is certainly not necessary for the existence of optimal controls (not even for the more general requirement of $\Phi(t_o, x_o)$ being closed) (see Neustadt, [1] and Cesari, [1], [2]).

One could say, that without the convexity condition, the attainable set in R^{n+1} : $\Phi(t_o, x_o)$ is not (in general) closed, because a limit x(t) of a sequence of trajectories $x_i(t)$, does not need to be an admissible trajectory (see again § 2, example 1). Physically, this does not make much sense, because any trajectory x(t) is experimentally known always with some error, and one could not distinguish the limit x(t) from a sufficiently near approximation $x_i(t)$. Therefore there is some interest in correcting this "fault" of the mathematical theory.

Gamkrelidze ([1][2]) gave a very elegant solution. Instead of considering the control equation (he takes the autonomous equation)

(II. 4. 1) $$x' = f(x, u)$$

with

(II. 4. 2) $$u(t) \text{ measurable}, \quad u(t) \in U \subset R^m,$$

he considers the equation

(II. 4. 3) $$x' = \sum_{i=1}^{n} p_i \, f(x, u_i) = g(x, p, v) ,$$

where the new control vector is $(p, v) = (p_1, \ldots, p_n, u_1, \ldots, u_n)$. This vector has to belong to the set

(II. 4. 4) $$(p_1, \ldots, p_n, u_1, \ldots, u_n) \in T^{n-1} \times U^n ,$$

where

(II. 4. 4') $$T^{n-1} = \{ p = (p_1, \ldots, p_n);\ p_1 + \ldots + p_n = 1,\ p_i \geqslant 0 \text{ for } i = 1, 2, \ldots, n \}$$

is a n-1 dimensional simplex, and

(II. 4. 4") $$U^n = U \times U \times \ldots \times U$$

is the cartesian power of the set U.

The set $U \in R^m$ is assumed to be compact and connected. The function f(x,u) is assumed to be continuous in (x, u), lipschitzian in x, and $x . f(x, u) \leqslant k (|x|^2 + 1)$. This last condition avoids trajectories going to infinity in finite time.

The set $C(x) = f(x, U)$ is then compact and connected, but of course not convex. Thus the existence theorems of Filippov and Ważewski are not applicable to the control system (II. 4. 1-2). On the other hand, the set

(II. 4. 5) $$g(x, T^{n-1}, U^n) = \left\{ \sum_{i=1}^{n} p_i \cdot f(x, u_i) \; ; \; p = (p_1, \ldots, p_n) \in T^{n-1}, \; u_i \in U \right\}$$

is the convex closure of the set $C(x)$, and the control system (II. 4. 3-4) satisfies the hypothesis of Filippov's theorem. Therefore, the time optimal control problem for the system (II. 4. 3-4) has always a solution (if x_1 is attainable at alla from x_o, for some time t_1, then there is a minimum time for attaining x_1). Gamkrelidze proved that the corresponding optimal solution $x(t)$ of system (II. 4. 3-4) can be uniformly approximated, as much as one wants, by solutions $x_i(t)$ of the system (II. 4. 1-2).

The key of the proof is the following : divide the interval $0 < t < T$, on which $x(t)$ is defined, into k subintervals $I_i^{(k)}$ $(i = 1, 2, , \ldots, k)$. Divide the segment $I_i^{(k)}$ into n measurable, non-intersecting sets $E_{ij}^{(k)}$, $j=1, 2, \ldots, n$, $I_i^{(k)} = \sum_{j=1}^{n} E_{ij}^{(k)}$, which satisfy the condition: measure $E_{ij}^{(k)} = \int_{I_i^{(k)}} p_j(t)\, dt$. ($p_j(t)$ is the function in (II. 4. 4) which corresponds to the solution $x(t)$ of (II. 4. 3).) Determine the control $u^{(k)}(t)$ in $0 \leqslant t \leqslant T$ using the relation $u^k(t) = u_j(t)$ for $t \in E_{ij}^{(k)}$, $j=1, 2, \ldots, n$; $i=1, 2, \ldots, k$. If for $k \to \infty$, the largest of the intervals $I_i^{(k)}$ tends to zero, then the sequence $u^{(k)}(t)$, $0 \leq t \leq T$, will be the desired one.

The functions $u_j(t)$ are called "base controls", the $p_j(t)$, the "weighting functions". From the described construction of the approximating sequence $u^{(k)}(t)$, it follows that the limit for $k \to \infty$ can be interpreted as a control "switching infinitely often" from one base control to another, in such a way that the time intervals spent on each base control are proportional to the corresponding weighting function. Therefore this kind of limit is sometimes referred to, as "chattering" or "sliding regime", and was used earlier by Filippov ([1]).

Another way to present this problem is given by the Polish school (Ważewski et. al.). Ważewski ([5], [6], [7]) considers systems defined as follows.

i) Let I be the compact interval $[0, a]$, r, M some positive constants and $h = \min (a, r/M)$. Let $B(r)$ denote the (n-dimensinnal) ball of center 0 and radius r, and K some compact set in R^m.

ii) Let $f(t, x, u)$ be continuous on $W = I \times B(r) \times K$ and $|f(t, x, u)| \leq M$.

iii) The set-valued function $U(t, x)$ is defined and Hausdorff continuous on $I \times B(r)$, where $U(t, x)$ is compact and $U(t, x) \subset K$.

The pair (f, U) is called a control system, and designated by $S(f, U)$. The set U is the control domain, and the image of this set through the function f, i. e. $C(t, x) = f(t, x, U(t, x))$ is called the orientor (function or field) associated to the control equation

(II. 4. 6) $x' = f(t, x, u)$,

(II. 4. 7) $u(t)$ measurable, $u(t) \in U(t, x)$.

Under these conditions, Wazewski defines :

Condition Y : $x(t)$ satisfies the condition $Y(f, U)$ if it is absolutely continuous on I, if $x(0) = 0$, and if there is a function $u(t)$, measurable on I and such that

$$x'(t) = f(t, x(t), u(t)), \quad u(t) \in U(t, x(t)) \quad (a.e. I) .$$

Condizion Z : $x(t)$ satisfies the condition $Z(f, U)$ if it is absolutely continuous on I, if $x(0) = 0$, and there are two sequences of functions , $x_i(t)$ absolutely continuous and $u_i(t)$ measurable on I, such that $x_i(0) = 0$, $u_i(t) \in U(t, x_i(t))$ (a.e. I) , $x_i'(t) - f(t, x_i(t), u_i(t)) \to 0$ (a.e. I) , $|x_i'(t)| \leqslant M$ (a.e. I) and $x_i(t) \to x(t)$ on I .

Definition : The functions $x(t)$ satisfying condition $Y(f, U)$ are the trajectories of the control system $S(f, U)$; similarly the functions $x(t)$ satisfying condition $Z(f, U)$ will be called "quasi-trajectories" of the control system $S(f, U)$.

Condition Y^* : $x(t)$ satisfies condition $Y^*(C)$ if it is absolutely continuous on I, $x(0) = 0$ and $x'(t) \in C(t, x(t))$ (a.e. I) .

Condition Z^* : $x(t)$ satisfies condition $Z^*(C)$ if there exists a sequence of absolutely continuous functions $x_i(t)$, defined on I , such that $x_i(0) = 0$, $|x_i'(t)| \leqslant M$, $x_i(t) \to x(t)$ on I , and distance $(x_i'(t), C(t, x_i(t))) \to 0$ (a.e. I)

Functions $x(t)$ which satisfy these last two conditions will be be called , respectively, trajectories and quasi-trajectories of the orientor field $C(t,x)$.

All these definitions correspond to trajectories and quasi-trajectories issuing from the origin ; it is obvious how to change the definition for the general case of initial condition $x(t_o) = x_o$.

According to Ważewski's theorem (seen in § 3) , condition $Y^*(C)$ is equivalent to condition $Y(f,U)$, and condition $Z^*(C)$ is equivalent to condition $Z(f,U)$. (this second statement is not so obvious as the first one, but it can be proved the same way)

Let $E(t,x)$ be the convex closure of the set $C(t,x)$. Then, assuming all the above mentioned hypothesis, the following properties hold :

a) The limit of a sequence of trajectories of $S(f,U)$ is a quasi-trajectory of $S(f,U)$.

b) Any trajectory of $S(f,U)$ is also a quasi-trajectory of $S(f,U)$.

c) If C is convex, any quasi-trajectory of the orientor field C is a trajectory of that field. (The limit of trajectories is a trajectory ; but observe that this last statement is weaker.)

d) Any trajectory of the orientor field E is a quasi-trajectory of the orientor field C. (If $x(t)$ satisfies $Y^*(E)$, it also satisfies $Z^*(C)$.)

This last result was proved by Ważewski ([5] , [6], [7] , [9] and is closely related to Gamkrelidze's theorem about sliding regimes. Gamkrelidze's assumptions are stronger (Lipschitz condition), but his result is also stronger : every trajectory of orientor field E is a limit of trajectories of the field C. Ważewski's result only says : every trajectory of the orientor field E is a quasi-trajectory of the field C (loosely speaking : a limit of approximate trajectories of C).

Sometimes it is useful to distinguish the quasi-trajectories, as we have defined them, calling them "weak quasi-trajectories", from the "strong quasi-trajectories" which are limits of trajectories. Then , under Gamkrelidze's assumptions, every trajectory of the field E is a strong quasi-trajectory of field C.

Turowicz ([1], [2], [3]) studied the set which is attainable by trajectories and the set which is attainable by quasi-trajectories . He gave a remarkable condition, which generalizes Gamkrelidze's theorem :

Assume $w(t, v)$ is a scalar function defined for $0 \leqslant t \leqslant h$, $0 \leqslant v < \infty$, non-negative, bounded and continuous, such that the differential equation $v'(t) = w(t, v(t))$ has $v(t) = 0$ as unique solution passing through the point (o, 0), on each interval $0 \leqslant t \leqslant \ell$, $0 < \ell \leqslant h$. Assume $|f(t, x, u) - f(t, \bar{x}, u)| \leqslant w(t, |x - \bar{x}|)$ for (t, x, u) $(t, \bar{x}, u)$ belonging to the domain of definition.

Assume also all the general hypothesis made before, and U = domain of control = constant, Then every quasi-trajectory of this control system is a limit of trajectories of it. In other words : every weak quasi-trajectory is a strong quasi-trajectory .

The question arises if there are weak quasi-trajectories which are not strong quasi-trajectories. That this is, indeed , possible, was shown by Pliś ([1]), with an example.

On the other hand, in the one-dimensional case (x = scalar), it was proved by Sedziwy ([1]) that every trajectory of the convex closure E, is a strong quasi-trajectory of the orientor field C. This is not the case, anymore, in two dimensions, even with $C(t, x) = C(x)$ independent of t, as the example of Pliś shows. (Every weak quasi-trajectory of field C is also a weak quasi -trajectory of field E, therefore by property (c) a trajectory of field E).

In some extent equivalent to the approach of Gamkrelidze and Ważewski, is the consideration of "generalized curves" (corresponding to quasi-trajectories). See Gambill ([1]), Leitman ([1]), Warga ([1], [2], [3]) McShane ([1], [2]) and Young ([1]).

Here we have studied properties of the attainable set (the "funnel" of solutions). In a recent paper, Pugh ([1]) poses the inverse problem : characterize those "funnels" $F(t, t_o, x_o)$ which are really attainable sets for a certain differential equation.

E.O. Roxin

III. Control systems defined abstractly.

§1. Generalized dynamical systems . The classical theory of differential equations is concerned to a great extent, with families of curves satisfying certain regularity conditions. The possible behaviour of such curves represent the behaviour of solutions of differential equations, and it is possible to develop a large part of the theory, disregarding the differential equations themselves. So, a general theory of "dynamical systems" or "topological dynamics" has been developed (for reference, see Gottschalk and Hedlund [1]).

This theory of dynamical systems starts with the axioms determining the behaviour of such systems , independently from the fact that they can (mostly) be described by differential equations. The main advantage of this approach lies in the fact that concepts like invariance, recurrence, stability etc. , are introduced in their greatest generality , showing their intrinsic nature.

For control systems, the same point of view can be adopted, leading to a theory of "generalized dynamical systems" . Compared with the classical dynamical systems, the main difference lies in the lack of uniqueness of the solution, given the initial data. Indeed, for a control system, this initial data determines a whole family of solutions, corresponding to the different admissible controls one may choose.

Among the different possibilities for starting such an abstract general theory, Bushaw's "polysystems" (Bushaw, [2]), where the basic elements correspond to the admissible trajectories of a control system, should be mentioned. Related is some work of Halkin ([2]).

Here we will present the theory of generalized dynamical systems based upon the notion of attainable set. This foundation of an abstract theory of control systems was first used by Barbashin ([1]), later thoroughly developed by Roxin ([4], [5], [6]).

The following notation will be used in the statement of the axioms which define the generalized dynamical systems.

i) $X = \{x, y, \dots\}$ will be the "state space", which we take to be the n-dimensional real space R^n, even if most of the properties hold for any complete, locally compact metric space.

ii) Subsets of X will be denoted by capital letters $A, B \dots$.

iii) $\rho(x, y)$ will denote the distance between points $x, y \in X$.

iv) $\rho(A, x) = \rho(x, A) = \inf \{\rho(x, y) \,;\, y \in A\}$ (distance between the point x and the set A).

v) $\beta(A, B) = \sup \{\rho(x, B) \,;\, x \in A\}$ ("deviation" of the set A from the set B).

vi) $\alpha(A, B) = \alpha(B, A) = \max \{\beta(A, B), \beta(B, A)\}$ (distance between the sets A, B in the Hausdorff pseudo-metric).

vii) $S_\varepsilon(A) = \{x \in X \,;\, \rho(x, A) \leqslant \varepsilon\}$ (ε-neighborhood of the set A).

A generalized dynamical system will be assumed to be determined by its "attainability function" $F(t, t_o, x_o)$, where $F \subset X$, $x_o \in X$, $t, t_o \in R$. The following axioms are assumed to hold:

(I) $F(t, t_o, x_o)$ is a closed nonempty subset of X, defined for every $x \in X$; $t, t_o \in R$.

(II) $F(t_o, t_o, x_o) = \{x_o\}$ for every $t_o \in R$, $x_o \in X$ (initial condition).

(III) For any $t_o \leqslant t_1 \leqslant t_2$, $x_o \in X$,

$$F(t_2, t_o, x_o) = \bigcup_{x_1 \in F(t_1, t_o, x_o)} F(t_2, t_1, x_1)$$

(semigroup property).

(IV) $x_1 \in F(t_1, t_o, x_o) \Longleftrightarrow x_o \in F(t_o, t_1, x_1)$.

(V) $F(t, t_o, x_o)$ is continuous in t: given t_1, t_o, x_o and $\varepsilon > 0$, there is $\delta > 0$ such that $\alpha(F(t, t_o, x_o), F(t_1, t_o, x_o)) < \varepsilon$ for all t such that $|t - t_1| < \delta$.

(VI) $F(t, t_o, x_o)$ is upper-semicontinuous in the triple (t, t_o, x_o): given t, t_o, x_o and $\varepsilon > 0$, there is $\delta > 0$ such that $\rho(x, x_o) < \delta$, $|\tau_o - t_o| < \delta$, $|\tau - t| < \delta$; imply $\beta(F(\tau, \tau_o, x), F(t, t_o, x_o)) < \varepsilon$.

This set of hypothesis is satisfied by a great number of control systems. It is also satisfied by solutions of differential equations which satisfy sufficient conditions for existence, but not for uniqueness of solutions. These generalized dynamical systems can, therefore, be regarded as a generalization in this last sense, of classical dynamical systems.

§ 2 . Properties of the generalized dynamical systems. Motions.

From the axioms given in the preceding section, it is not difficult to prove the following properties of generalized dynamical systems .

i) If $A \subset X$, $T_o \subset R$, $T \subset R$ are compact non-empty sets, then the attainability function

$$F(T, T_o, A) = \bigcup_{\substack{t \in T \\ t_o \in T_o \\ x \in A}} F(t, t_o, x)$$

is upper semicontinuous in (T, T_o, A) : given (fixed) sets T, T_o, A, and $\varepsilon > 0$, there is $\delta > 0$ such that for any T', T_o', A' with

$$\rho(T, T') < \delta, \ \rho(T_o, T_o') < \delta, \ \rho(A, A') < \delta,$$

the relation

$$\rho(F(T', T_o', A'), \ F(T, T_o, A)) < \varepsilon$$

follows.

ii) The set $F(t, t_o, x_o)$ is not only closed (axiom I) but also compact. This follows from the continuity (axiom V) .

iii) For compact sets T, T_o, A, the set $F(T, T_o, A)$ is also compact.

It is interesting to note (and this is a difference with upper semicontinuous contingent equations) , that the attainable set $F(t, t_o, x_o)$ is not necessarily connected. To see this it is sufficient to consider the following example :

Define the generalized dynamical system by the "differential equations" for $|x| > |y| \ldots \quad \begin{cases} x' = -\operatorname{sgn} x \\ y' = 0 \end{cases}$

for $|x| < |y| \neq 0 \quad \begin{cases} x' = 0 \\ y' = \operatorname{sgn} y \end{cases}$

for $x = y = 0 \quad \begin{cases} \text{the right derivatives} & x' = 0 \\ & y' = \pm 1 \text{ (both values)} \\ & x' = \pm 1 \text{ (both values)} \\ \text{the left derivatives} & y' = 0 \end{cases}$

for $|x| = |y| \neq 0 \quad : \begin{cases} x' = 0, \pm 1 \\ y' = 0, \pm 1. \end{cases}$

This is really a contingent equation, but it is not included in the theory of Marchaud and Zaremba (seen in the preceding chapter), because at $x = y = 0$ the set of possible derivatives $C(t, x)$ is not convex, not upper semicontinuous and even different for right and left derivatives. It is therefore not surprising that the set $F(t, 0, 0)$, which can be obtained very easily by direct considerations (see figure 3.1), is $(0, \pm t)$ for $t > 0$, $(\pm t, 0)$ for $t < 0$, therefore not connected. On the other hand, all the axioms I - VI given in the preceding paragraph are satisfied, and this is a generalized dynamical system.

The decisive step, by which generalized dynamical systems defined by axioms I - VI of the preceding paragraph, behave according to what should be expected, is the existence of "<u>trajectories</u>" or "<u>motions</u>".

We define a <u>motion</u> of the generalized dynamical system determined by $F(t, t_o, x_o)$, as a function $u : [t_o, t_1] \rightarrow X$, such that $t_o \leqslant t_a \leqslant t_b \leqslant t_1$ implies

(III. 2. 1) $\qquad u(t_b) \in F(t_b, t_a, u(t_a))$.

Theorem I : if $x_1 \in F(t_1, t_o, x_o)$, then there exists a motion $u(t)$, such that $u(t_o) = x_o$, $u(t_1) = x_1$.

To prove this , we assume, for simplicity, $t_o = 0, t_1 = 1$. A motion satisfying the desired conditions can be constructed the following way :

For $t = 1/2, 1/4, 3/4, 1/8, 3/8, \ldots, p/2^q, \ldots$, the values of $u(t)$ can be chosen successively such that

$$u(p/2^q) \in F(p/2^q, (p-1)/2^q, u((p-1)/2^q)) \cap F(p/2^q, (p+1)/2^q, u((p+1)/2^q)).$$

This defines $u(t)$ for all values of t which are binary fractions . For all other values of t,

$$\bigcap_{\substack{t' = \\ = \text{binary fraction}}} F(t, t', u(t')) = K(t)$$

is non-empty , because it is the intersection of compact sets with the finite intersection property (it is, indeed, a single point) . Taking $u(t) \in K(t)$, the motion is obtained in the interval $t \in [0, 1]$ and satisfies the desired property. Indeed, for any $t_a < t_b$, there is a binary fraction $t \in (t_a, t_b)$ and

$$u(t_b) \in F(t_b, t, u(t)) \subset F(t_b, t_a, u(t))$$

(see figure 3.2) .

The main tool for applications is the following

Theorem II (Barbashin [1]) : If $u_i(t)$ are motions of a given generalized dynamical system, which are defined in the interval $a \leqslant t \leqslant b$, and if for some $a \leqslant t_o \leqslant b$, $u_i(t_o) \to x_o$ for $i \to \infty$, then there is some subsequence $u_{i_j}(t)$ which converges uniformly in $[a, b]$ to some motion $u_o(t)$.

The values $u_i(t_o)$ belong to a certain compact set $S \subset X$, and hence for every $t \in [a, b]$, $u_i(t) \in S_t = F(t, t_o, S)$ which is also compact.

Taking a countable dense subset $\{t_1, t_2, t_3, \dots\} \in [a, b]$, it is possible to choose subsequences converging at the first n values $t_1, t_2, \dots, t_n$. By the classical diagonal procedure (as in the proof of Arzelá's theorem about equicontinuous families of functions), a subsequence is obtained which converges at all values $t_1, t_2, t_3, \dots$. We will assume that this subsequence is the original sequence $u_i(t)$ (for simplicity). At the remaining values of $t \in [a, b]$, we define the limit function $u_o(t)$ as in the preceding theorem.

It remains to show that, indeed, $u_i(t) \to u_o(t)$ and this even uniformly. Assuming the contrary, there is some subsequence $u_{i_j}(t)$ and corresponding values $t_j \in [a, b]$, $(j=1, 2, 3, \dots)$, such that $t_j \to \theta$ and $u_{i_j}(t_j) \to y \neq u_o(\theta)$ for $j \to \infty$. Take any value τ belonging to the dense subset, where the sequence $u_i(t)$ converges by hypothesis. Then $u_i(\tau) \to u_o(\tau)$, and by upper semi-continuity, $u_i(t_j) \in F(t_j, \tau, u_i(\tau))$ implies

$$y \in F(\theta, \tau, u_o(\tau)) .$$

This relation holds for any τ belonging to the dense subset, therefore taking τ variable, $\tau \to \theta$, we obtain finally

$$y \in \lim F(\theta, \tau, u_o(\tau)) = F(\theta, \theta, u_o(\theta)) = u_o(\theta),$$

contrary to our assumption. Therefore the theorem is proved.

§ 3. Relation with contingent equations.

It is quite obvious that contingent equations, which are the intrinsic formulation of control equations, must be closely related to the generalized dynamical systems defined abstractly. This matter has been investigated by Roxin ([7]), on the basis of the theory of Marchaud ([1], [2], [3]) and Zaremba ([1], [2]) and in a recent paper of Turowicz ([2]).

Theorem I: Given the contingent equation

$$x' \in C(t,x), \tag{III.3.1}$$

where $C(t,x) \subset X$ is defined in $X \times R$, is compact and convex for each (t,x), upper semicontinuous in (t,x), and satisfies the condition

$$\sup \left\{ |y| \; ; \; y \in C(t,x) \right\} \leqslant k \quad (1 + |x|) \tag{III.3.2}$$

or other similar condition which rules out a "finite escape time" (the possibility that for some solution $x(t)$, $\lim |x(t)| = \infty$ for $t \to T$ finite), and if we define

$$\begin{cases} x_1 \in F(t_1, t_o, x_o) \Leftrightarrow \text{there exists a trajectory } x = u(t) \\ \text{of (III.3.1) such that } u(t_o) = x_o,\ u(t_1) = x_1, \end{cases} \tag{III.3.3}$$

then this $F(t_1, t_o, x_o)$ determines a generalized dynamical system.

It is quite easy to check the fact that the axioms I - VI are satisfied.

Definition: Given the generalized dynamical system $F(t, t_o, x_o)$ and a point (t_o, x_o), the contingent of the generalized dynamical system at that point, designated by $D^* F(t, t_o, x_o)$, is defined to be the union of all the contingents (at that point) of all the motions $u(t)$ of that system, which pass through (t_o, x_o) :

(III. 3. 4) $$D^* F(t, t_o, x_o) = \bigcup_{u(t_o) = x_o} D^* u(t) .$$

With this definition, it makes sense to write a "contingent equation for a generalized dynamical system"

(III. 3. 5) $$D^* F(t, t_o, x_o) = C(t_o, x_o) ,$$

$C(t_o, x_o)$ being given . The questions of existence and uniqueness of solutions arise, as in the classical theory of ordinary differential equations. In the following we will only give some interesting results, and refer the reader to the literature cited above.

Existence Theorem II : If the variable set $C(t_o, x_o) \subset X$ is continuous in (t_o, x_o) and satisfies the other conditions of the preceding theorem, then there exists a generalized dynamical system $F(t, t_o, x_o)$ satisfying equation (III. 3. 5)

Uniqueness Theorem III. If $C(t_o, x_o)$ satisfies the conditions of the existence theorem , and moreover the "Lipschitz condition"

(III. 3. 6) $$\alpha(C(t, x_2), C(t, x_1)) \leqslant K. |x_2 - x_1| ,$$

then the solution $F(t, t_o, x_o)$ of (III. 3. 5) is unique.

It is interesting to note that given a generalized dynamical system $F_1(t, t_o, x_o)$, its contingent (according to the above definition) defines a set function

$$C(t_o, x_o) = D^* F_1(t, t_o, x_o) .$$

We can write the general contingent equation

(III. 3.7) $\quad D^{*} F(t, t_o, x_o) = C(t_o, x_o)$,

of which F_1 is, of course, a solution. But we can try to obtain the solution of (III. 3.7) , determining all possible trajectories x(t) of the (ordinary) contingent equation (III. 3.1) , and applying the criterion of the first theorem of this paragraph (take $x_1 \in F(t_1, t_o, x_o)$ if there is a trajectory of (III. 3.1) going from (t_o, x_o) to (t_1, x_1)). But even if the set $C(t_o, x_o)$ satisfies the conditions of theorem II, we might by this procedure obtain a generalized dynamical system $F_2(t, t_o, x_o)$ different from F_1. This is due, of course, because there might not be uniqueness of the solutions, and is shown in the following example.

Let the motions of the generalized dynamical system be:

$$x = u(t) = (t + \text{const.})^3 .$$

Then

$$F_1(t, t_o, x_o) = \left\{ (t - t_o + \sqrt[3]{x_o})^3 \right\} \text{(the set of one point)}$$

The contingent equation (III. 3.1) becomes the ordinary differential equation

$$x' = 3\, x^{2/3} ,$$

and , for $t > 0$,

$$F_2(t, 0, 0) = [0, t^3]$$

is the set $0 \leqslant x \leqslant t^3$.

If this situation occurs, it seems reasonable to call "maximal solution " that generalized dynamical system (like F_2) , which is obtained by all the trajectories of the corresponding ordinary contingent equation.

E.O. Roxin

IV . Control systems and calculus of variations.

§ 1. The optimal control problem. We have already mentioned that, given a control system by its differential equation

(IV.1.1) $$x' = f(t, x, u)$$

with the condition

(IV.1.2) $u \in K$ = class of admissible controls,

one of the basic problems is to find the "attainable set" $F(t, t_o, x_o)$. On the other hand, it is obviously hopeless to try to solve this problem in general, because even in the "simple" case when there is no control at all, and (IV.1.1) is an ordinary differential equation, $F(t, t_o, x_o)$, which is then the general solution, cannot be found explicitly in general.

If there are given, besides the control equations (IV.1.1-2), the boundary conditions

(IV.1.3) $$x(t_o) = x_o, \quad x(t_1) = x_1,$$

it may happen that $(t_1, x_1) \in F(t_1, t_o, x_o)$ ("is attainable from $(t_o x_o)$") or not. In the first case, there are in general more than one control $u(t)$ such that the corresponding solution $x(t)$ satisfies (IV.1.3) (a control which "transfers (t_o, x_o) to (t_1, x_1)"). In that case, it makes sense to ask for the "best" control to choose from all possible ones. This means that there has to be given a certain "optimality criterion" for the controls $u(t)$ or for the corresponding solutions $x(t)$.

According to the necessities of applications, the boundary conditions (IV. 1.3) are generally relaxed as follows :

(IV. 1.4) given, t_o, x_o, $G \subset R^{n+1}$

satisfy $x(t_o) = x_o, x(t_1) = x_1, (t_1, x_1) \in G$.

An important case of (IV. 1.4) is when the set G is the half-line $t \geq t_o$, $x = x_1$. In this case the boundary conditions can remain the (IV. 1.3) but, well understood, t_o, x_o, x_1 fixed , t_1 arbitrary.

The optimality criterion is generally to minimize a certain functional

(IV. 1.5) $$J = \varphi(t_1, x_1) + \int_{t_o}^{t_1} f^o(t, x, u)\, dt$$

This corresponds to the classical "Bolza problem" in the calculus of variations. The case with $\varphi(t_1, x_1) = 0$ is called "Lagrange problem" ; when $f^o(t, x, u) = 0$, we have the "Mayer problem". As a reference for the classical calculus of variations, the reader may see Bliss ([1]) or Carathéodory ([1]) .

It is well known that the problems of Bolza, Mayer and Lagrange are equivalent, because they can be transformed in each other easily. Therefore we will refer , as is usual, to the Lagrange problem

(IV. 1.6) $$J = \int_{t_o}^{t_1} f^o(t, x, u)\, dt = \text{minimum}$$

$f^o(t, x, u)$ is supposed to be continuous and non-negative. A very important special case is $f^o = 1$, for then $J = t_1 - t_o$ and the problem, known as the "time optimal problem", consists in finding the "time optimal control " which transfers x_o into x_1 in least time. This type of control was first considered (in this kind of problem) by Bushaw ([1]).

As for the class of admissible controls K, it is generally assumed

(IV.1.7) $$K = \left\{ u : R \rightarrow R^m \text{ measurable } ; u(t) \in U \subset R^m \right\}$$

with U compact. Sometimes $U = U(t, x)$ is considered.

§ 2 . Pontryagin's Maximum Principle.

Pontryagin, Boltyanskii and Gamkrelidze ([1] , andtogether with Mischchenko [1]) developed a necessary contidion for optimality, applicable to problems of optimal control, known as Pontryagin's Maximum Principle, which generalizes very nicely some classical results. For the proof of this principle, the reader is referred to the above mentioned literature, and here we will only give the formulation of this principle.

We first consider the problem of an autonomous system

(IV.2.1) $$x^{i\,\prime} = f^i(x^1, \ldots, x^n, u) \quad (i=1, 2, \ldots, n)$$

where $u \in R^m$ and the functions f^i are assumed continuous with respect to all variables $x^1, x^2, \ldots, x^n, u$ and continuously differentiable with respect to $x^1, x^2, \ldots, x^n$. The class of admissible controls is

(IV.2.2) $$K = \left\{ u \text{ measurable}; u(t) \in U \subset R^m \right\} ,$$

The boundary conditions are

(IV. 2. 3) $\quad x^i(t_o) = x^i_o \;,\; x^i(t_1) = x^i_1 \;;\; t_o, x^i_o, x^i_1$ given,

t_1 = arbitrary $(i=1, 2, 3, \dots, n)$.

The optimality criterion is

(IV. 2. 4) $$J = \int_{t_o}^{t_1} f^o(x^1(t), \dots, x^n(t), u(t))\, dt = \text{minimum},$$

with f^o non-negative and satisfying similar conditions as $f^i(x^1, \dots, u)$.

The problem is transformed into a Mayer problem, by taking one more component x^o of the x-vector, such that

(IV. 2. 5) $$x^{o\prime} = f^o(x^1, \dots, x^n, u) \;, \quad x^o(t_o) = 0 \;.$$

The optimality criterion is, then,

(IV. 2. 6) $$x^o(t_1) = \text{minimum} \;.$$

<u>Theorem A:</u> Let $u(t)$ be an admissible control, $x^i(t)$ $(i=0, 1, \dots, n)$ be the corresponding trajectory of (IV. 2. 1-5), satisfying the boundary conditions (IV. 2. 3) In order that $u(t)$ and $x^i(t)$ be optimal for $t_o \leqslant t \leqslant t_1$, it is necessary that there exists a nonzero, absolutely continuous vector function $p(t) = \{p_o(t), p_1(t), \dots, p_n(t)\}$ such that:

1) $x^i(t)$, $p_i(t)$ and $u(t)$ satisfy the Hamiltonian system

(IV. 2. 7) $$\begin{cases} \dfrac{dx^i}{dt} = \dfrac{\partial H}{\partial p_i} \\[2ex] \dfrac{dp_i}{dt} = -\dfrac{\partial H}{\partial x^i} \end{cases} \qquad (i = 0, 1, 2, \dots, n) \;,$$

where

(IV. 2. 8) $$H(p, x, u) = \sum_{i=0}^{n} p_i \, f^i(x, u) \;.$$

2) For almost all t, $t_o \leqslant t \leqslant t_1$, the function $H(p(t), x(t), u)$ achieves a maximum at $u = u(t)$:

(IV. 2. 9) $$H(p(t), x(t), u(t)) = M(p(t), x(t)) \quad \text{a.e.} ,$$

M being the maximum of H :

(IV. 2. 10) $$M(p, x) = \sup \left\{ H(p, x, u) \; ; \; u \in U \right\} .$$

3) At the initial time t_o

(IV. 2. 11) $$p_o(t_o) \leqslant 0 , \quad M(p(t_o), x(t_o)) = 0 .$$

If the variables $p(t)$, $x(t)$, $u(t)$ satisfy conditions (1) and (2), then the functions of time $p_o(t)$ and $M(p(t), x(t))$ are constant, so that relations (IV. 2. 11) are verified at any $t \in [t_o, t_1]$.

The vector $p(t)$ is called "adjoint vector" or "co-vector".

In the case of $f^o = 1$, which defines the "time optimal" control the above conditions give :

1) The nonzero, absolutely continuous vector $p(t) = \left\{ p_1(t), \ldots \right.$ $\left. \ldots, p_n(t) \right\}$ satisfies the Hamiltonian system

(IV. 2. 12) $$\frac{dx^i}{dt} = \frac{\partial H}{\partial p_i} , \quad \frac{dp_i}{dt} = - \frac{\partial H}{\partial x^i} \quad (i=1, 2, \ldots, n) ,$$

where

(IV. 2. 13) $$H(p, x, u) = \sum_{i=1}^{n} p_i f^i(x, u) .$$

2) Almost everywhere in $[t_o, t_1]$,

(IV. 2. 14) $$H(p(t), x(t), u(t)) = M(p(t), x(t)) = \sup \left\{ H(p(t), x(t), u) \; ; \; u \in U \right\} .$$

3) At the initial time t_o :

(IV. 2. 15) $\quad M(p(t_o), \; x(t_o)) \geqslant 0 \; .$

The non-autonomous case is easily reduced to an autonomous case of higher dimension. The result is :

Theorem B : Given the problem stated as before ((IV. 2.1) to (IV. 2. 4)) , but now with $f^i = f^i(t, x, u)$ $(i=0, 1, 2, \ldots, n)$, in order that $u(t)$, $x(t)$ be optimal in the interval $[t_o, t_1]$, it is necessary that there exists an absolutely continuous, nonzero vector $p(t) = \{ p_o(t), \; p_1(t), \ldots, p_n(t) \}$ such that :

1) $p(t)$, $x(t)$, $u(t)$ satisfy the Hamiltonian system (IV. 2. 7) with (IV. 2. 8) .

2) (IV. 2. 9) is also satisfied a.e., with (IV. 2. 10) .

3)

(IV. 2. 16) $\quad p_o(t) = \text{const} \leqslant 0 \; ; \quad M(p(t_1)\, , \, x(t_1)\, , \, t_1) = 0 \, .$

Here , $M(p(t), \; x(t), \; t)$ is not constant ; indeed ,

$$M(p(t), x(t), t) = \int_{t_1}^{t} \sum_{i=0}^{n} \frac{\partial f^i(x(t), u(t), t)}{\partial t} \; p_i(t) \, dt .$$

Finally , we have to consider more general type of boundary conditions.

Theorem C: Assume the problem is similar to the one of theorem A or B, but the boundary conditions at the right are :

(IV. 2. 17) $\quad x(t_1) \in G \subset R^m \quad \text{or} \quad (t_1, x_1) \in G \subset R^{n+1} \; .$

where G is a given set . Then, the point x_1 (respectively (t_1, x_1)) must belong to the boundary of G . Assume at (t_1, x_1) , G has a tangent hyperplane T (or a tangent manifold contained in T). Then, the vector $p(t_1)$ must be orthogonal to the tangent plane T. This condition is referred to, as the "transversality condition".

Theorem D: If the problem is similar to the previous ones but with fixed time, i.e. t_1 is fixed, then all the conditions of theorem B hold, except the last one : $M(p(t_1), x(t_1), t_1) = 0$, which is dropped; it is, indeed, replaced by the condition: t_1 given.

§3. Relation with the classical Weierstrass-function.

If the control set U is open, or, more generally, if the optimal control u(t) takes only interior values of U, then Pontryagin's Maximum Principle coincides with the classical necessary Weierstrass condition for a minimum.

Let us consider the autonomous case of theorem A of the preceding paragraph. In order to apply the classical Weierstrass condition, one has to transform the Lagrange problem with the so called "multiplier rule".

The integral to be minimized is

(IV. 3. 1) $$J = \int_{t_o}^{t_1} f^o(x, u)\; dt$$

and the "side conditions" are

(IV. 3. 2) $$\Phi_i = f^i(x, u) - \frac{dx^i}{dt} = 0, \quad (i=1, 2, \ldots, n) .$$

Therefore one has to consider the function

(IV. 3. 3) $F = \lambda_o f^o + \lambda_i \Phi_i$,

where the λ_i 's are the undetermined multipliers . We identify these with the components of the vector $-p$, so that

(IV. 3; 4) $F = -p_o f^o - \sum_{i=1}^{n} p_i \left(f^i(x,u) - \frac{dx^i}{dt}\right) = -H(p,x,u) + \sum_{i=1}^{n} p_i \frac{dx^i}{dt}$.

The Weierstrass condition is formulated as $E(t, x, x', \bar{x}') \geqslant 0$, where $x' = x'(t)$ is the derivative of the (optimal) trajectory, $\bar{x}'$ is any other (admissible) value of the derivative, and E is the "Weierstrass function"

(IV. 3. 5) $E = F(t,x,\bar{x}') - F(t,x,x') - (\bar{x}'-x') F_{x'}(t,x,x')$.

In order to take into account the control parameters u^i , we "enlarge" the vector x by writing

(IV. 3. 6) $\frac{dx^{j+n}}{dt} = u^j \quad (j = 1, 2, \ldots, m)$.

We then obtain,

$$E = \left[-H(p(t), x(t), \bar{u}) + \sum p_i(t) f^i(x(t), \bar{u})\right] - $$
$$- \left[-H(p(t), x(t), u(t)) + \sum p_i(t) f^i(x(t), u(t))\right] +$$
$$+ \sum (\bar{u}^j - u^j) \frac{\partial H(p(t), x(t), u(t))}{\partial u^j} - \sum (f^i(x(t), \bar{u}) -$$
$$- f^i(x(t), u(t))) p_i(t) =$$

$$= H(p(t), x(t), u(t)) - H(p(t), x(t), \bar{u}) +$$
$$+ \sum (\bar{u}^j - u^j) \frac{\partial H(p(t), x(t), u(t))}{\partial u^j}$$

Now, at any <u>interior</u> point of U, the derivatives

$$\frac{\partial H(p(t), x(t), u(t))}{\partial u^j}$$

are zero (H takes precisely its maximum value), and this both from the Maximum Principle and from the classical theory. Therefore, for interior points of U, the condition $E \geqslant 0$ is equivalent to

$$H(p(t), x(t), u(t)) \geqslant H(p(t), x(t), \bar{u})$$

for every $\bar{u} \in U$, which is precisely the Maximum Principle.

V. Boundary controls.

§ 1. Boundary motions and boundary controls. Consider the control equation defined by

(V. 1. 1) $x' = f(t, x, u) \qquad (x \in R^n, \ u \in R^m)$,

where f satisfies for example the conditions specified in Pontryagin's Maximum Principle, with the family of admissible controls

(V. 1. 2) $K = \{u(t) \text{ measurable} \ ; \ u(t) \in U = U(t, x)\}$,

or, equivalently, the contingent equation

(V. 1. 3) $x' \in C(t, x)$

with

(V. 1. 4) $C(t, x) = f(t, x, U(t, x))$.

We shall call "boundary controls" those control functions u(t) , for which

(V. 1. 5) $f(t, x, u(t)) \in \partial f(t, x, U(t, x)) = \partial C(t, x)$,

where ∂A = boundary of A . Attention should be called on the fact that not necessarily

$$u(t) \in \partial U(t, x) ,$$

which nevertheless is true in the important case, in which equation (V. 1. 1) is linear in the control u .

The importance of boundary controls is shown by Pontryagin's principle, where the optimal control $u_o(t)$ satisfies almost everywhere, for some vector $p = p(t)$,

$$p. f(t, x, u_o(t)) = \max \{p. f(t, x, u) ; \ u \in U\},$$

and therefore $u_o(t)$ is almost everywhere a boundary control.

Let (t_o, x_o) be fixed, and let

(V.1.6) $$F(t, t_o, x_o)$$

be the attainable set from (t_o, x_o), for a given control system of the type (V.1.1-2). Then a certain motion $x(t)$ of the control system will be called <u>"boundary motion"</u> in $[t_o, t_1]$ if, for all values of $t \in [t_o, t_1]$,

(V.1.7) $$x(t) \in \partial F(t, t_o, x_o) .$$

Boundary motions represent optimal motions in some sense ; for example in the time optimal control problem, if (t_1, x_1) is attainable from (t_o, x_o) and t_1 is minimal (compared to all other (t, x_1) which are attainable from (t_o, x_o)), then $(t_1, x_1) \in \partial \Phi(t_o, x_o)$. If every subarc of the motion $x(t)$ is time-optimal, the whole motion $x(t)$ will belong to the boundary of the attainable set.

About existence of boundary motions, Fukuhara gave a theorem for the case of differential equations without uniqueness, (see Fukuhara [1], [2] and Kamke [1]), extended by Marchaud ([3]) and Zaremba ([2]) to contingent equations. In the following we give this theorem for generalized dynamical systems.

<u>Theorem I</u>: If, for a generalized dynamical system F,

(V.1.8) $$x_1 \in \partial F(t_1, t_o, x_o) , \qquad (t_1 \geq t_o) ,$$

Then there is a boundary motion $x(t)$ joining $x(t_o) = x_o$ with $x(t_1) = x_1$:

(V.1.9) $$x(t) \in \partial F(t, t_o, x_o) \qquad \text{for } t_o \leq t \leq t_1 .$$

To prove this theorem, we will construct the boundary control $x(t)$ as limit of a sequence of motions $y_i(t)$,

For every set $t_2, t_3, \ldots, t_n$, let $t_{n_2}, t_{n_3}, \ldots, t_{n_n}$ be the same set but ordered : $t_{n_2} \leqslant t_{n_3} \leqslant \ldots \leqslant t_{n_n}$. Let also, for every n, $t_{n_1} = t_o$, $t_{n_{n+1}} = t_1$. Then a corresponding sequence $x_{n_1} = x_o, x_{n_2}, x_{n_3}, \ldots, x_{n_{n+1}} = x_1$ can be found, such that $x_{n_i} \in \partial F(t_{n_i}, t_o, x_o)$ and

$$x_{n_{i+1}} \in F(t_{n_{i+1}}, t_{n_i}, x_{n_i}) \qquad (i = 1, 2, \ldots, n).$$

Indeed, to find this sequence, the above given construction has to be carried out, first for $t_{n_n} \in [t_o, t_1]$, then for $t_{n_{n-1}} \in [t_o, t_{n_n}]$, and so on (see figure V.1). Therefore a motion $y_n(t)$ exists, such that

$$y_n(t_o) = x_o, \quad y_n(t_{n_i}) = x_{n_i}, \quad y(t_1) = x_1.$$

These points x_{n_i} belong to the boundary of $F(t_{n_i}, t_o, x_o)$.

By the thorem of Barbashin given in chapter III, from the motions $y_n(t)$ we can select a converging subsequence. This subsequence converges to some motion $x(t)$. Obviously, $x(t_o) = x_o$, $x(t_1) = x_1$. For every value of t belonging to the dense subset $\{t_i\}$, $x(t_i) \in \partial F(t_i, t_o, x_o)$, because the same happens to all the $y_n(t)$ (with the possible exception of a finite number of them). As the boundary is a closed set,

$$x(t) \in \partial F(t, t_o, x_o) \qquad (t_o \leqslant t \leqslant t_1)$$

results from the fact that $\{t_i\}$ is dense in $[t_o, t_1]$.

This theorem says nothing about the possible existence of other motions joining (t_o, x_o) with (t_1, x_1), which are not boundary

$i = (1, 2, 3, \dots)$.

Given any t_2, $t_o \leq t_2 \leq t_1$, by axiom III of the generalized dynamical systems,

$$F(t_2, t_o, x_o) \cap F(t_2, t_1, x_1) \neq \emptyset .$$

Moreover,

$$F(t_2, t_1, x_1) \cap \partial F(t_2, t_o, x_o) \neq \emptyset .$$

Indeed, if $F(t_2, t_1, x_1)$ would not intersect $\partial F(t_2, t_o, x_o)$, it would belong to the interior of $F(t_2, t_o, x_o)$. As both sets $F(t_2, t_i, x_i)$ are closed, there would be a neighborhood of $F(t_2, t_1, x_1)$ also belonging to $F(t_2, t_o, x)$. By upper semicontinuity of $F(t_2, t_1, x)$ in x (axiom VI), there is some neighborhood of x_1, V, such that for every $x \in V$,

$$F(t_2, t_1, x) \subset F(t_2, t_o, x_o).$$

But then (t_1, x) is attainable from (t_o, x_o) because for any $z \in F(t_2, t_1, x)$,

$$z \in F(t_2, t_o, x_o) \quad \text{and} \quad x \in F(t_1, t_2, z) , \quad t_o \leq t_2 \leq t_1 .$$

(Here we apply axioms III and IV.) This contradicts the hypothesis that x_1 belongs to the boundary of $F(t_1, t_o, x_o)$, because in any neighborhood of x_1 there are points (t_1, x) not attainable from (t_o, x_o).

We have so proved the existence of some point $x_{2,1} \in \partial F(t_2, t_o, x_o)$ such that $x_1 \in F(t_1, t_2, x_{2,1})$. By the existence theorem of motions of a generalized dynamical system, there is some motion $y_1(t)$, such that $y_1(t_o) = x_o$, $y_1(t_2) = x_{2,1}$, $y_1(t_1) = x_1$.

Take a sequence $t_2, t_3, \dots$, which is dense in the interval $[t_o, t_1]$.

motions,. This possibility is ruled out by the next theorem, in which some additional assumptions have to be made .

Theorem II. (Roxin, [2]) ; If , in (V.1.2) the control set $U(t,x) = U(t)$ does not depend on x, and if in (V.1.1) $f(t,x,u)$ is such that for every admissible control $u(t) \in K$, the solution $x(t)$ of (V.1.1) exists, is unique, does not escape to infinity in finite time, and depends continuously on the initial data, then : if $x_1 \in \partial F(t_1, t_o, x_o)$, for every motion $x(t)$ joining $x(t_o) = x_o$ with $x(t_1) = x_1$, $x(t) \in \partial F(t, t_o, x_o)$ holds for every $t \in [t_o, t_1]$. In other words, all motions joining (t_o, x_o) with (t_1, x_1) are boundary motions.

According to the hypothesis, given the values t_1, t_2 and a motion $x(t)$ with $x(t_1) = x_1$, $x(t_2) = x_2$, this motion corresponds necessarily to some control $u(t)$. For this particular control, (V.1.1) is an ordinary differential equation, and the uniqueness and continuous dependence (both ways) between $x_u(t_1)$ and $x_u(t_2)$ (where $x_u(t)$ denotes any solution corresponding to the particular control $u(t)$), implies that this correspondence is topological.

Therefore, if we assume $t_o \leq t_2 \leq t_1$, it is easy to show that for any control $x(t)$ with $x(t_o) = x_o$, $x(t_1) = x_1$, denoting $x_2 = x(t_2)$, the relation $x_2 \in \partial F(t_2, t_o, x_o)$ holds. Indeed, if there were a neighborhood V_2 of x_2 contained in $F(t_2, t_o, x_o)$, there would be also a neighborhood V_1 of x_1 such that for every $y_1 \in V_1$, the corresponding solution $x_u(t)$ (i.e. with $x_u(t_1)\, y_1)$), satisfies $x_u(t_2) = y_2 \in V_2$. But then $y_1 \in F(t_1, t_o, x_o)$ for every $y_1 \in V_1$, contrary to the hypothesis (see figure V.2) .

This property is an important step in the proof of Pontryagin's principle. To show that under the general assumptions of theorem I, it is not always true (therefore some additional restrictions, like those of theorem II, have to be made), a counterexample of G. Haynes and H. Hermes is given next.

Let x be a scalar variable,

$$x' = f(x, u) = u \exp\left(-\frac{u}{\sqrt{1-x^2}}\right)$$

the differential equation, and

$$0 \leqslant u(t) \leqslant 1$$

the control set. (Note that $f(x, u)$ does not satisfy the Lipschitz condition).

Let $x_o = x(0) = 0$, Then, for $t \geqslant 0$, the control $u(t) = 0$ gives the lower boundary of attainable set, $x_{min} = 0$. The upper boundary is obtained for x' = maximum, which gives $u = \sqrt{1-x^2}$. Therefore we have to integrate the equation.

$$x' = \frac{1}{e}\sqrt{1 - x^2},$$

whose solution is

$$x(t) = \begin{cases} \sin \frac{t}{e} & \text{for } 0 \leqslant t \leqslant \frac{\pi e}{2} \\ 1 & \text{for } t \geqslant \frac{\pi e}{2} \end{cases}.$$

The attainable set is shown in figure V.3. From the differential equation it is seen that the control $u_1(t) = 0$ for $0 \leqslant t < t_o$, $u_1(t) = \sqrt{1 - x^2}$ for $t_o \leqslant t$, gives the solution

$$x_1(t) = \begin{cases} 0 & \text{for} \quad 0 \leqslant t \leqslant t_o \\ \sin \frac{t}{e} & \text{for} \quad t_o \leqslant t \leqslant t_o + \frac{\pi e}{2} \\ 1 & \text{for} \quad t_o + \frac{\pi e}{2} \leqslant t . \end{cases}$$

Hence, $x_1 = 1$ is attainable, for $t = t_o + \frac{\pi e}{2}$ by a boundary motion and by other non-boundary motions.

§ 2. Extremal solutions for linear systems. Here we will consider control equations of the type

(V. 2. 1) $$x' = A(t) \; x + g(t, u) , \qquad (x = n\text{-vector}, \; u = m\text{-vector})$$

with control domain

(V. 2. 2) $$U = \text{const.} \subset R^m ,$$

and admissible controls

(V. 2. 3) $$\mathcal{K} = \{ u \text{ measurable} ; \; u(t) \in U \} .$$

The set U is assumed to be compact; $A(t)$ is an $n \times n$ matrix, whose elements are Lebesgue integrable on each finite interval. The function $g(t, u) : R \times U \rightarrow R^n$ is assumed to be continuous in u for each t and Lebesgue measurable in t for each u. Moreover,

(V. 2. 4) $$|g(t, u)| \leqslant m(t) \qquad \text{for each } t \in R, \; u \in U,$$

where $m(t)$ is Lebesgue integrable on each finite interval of R.

Under these conditions, the admissible controls can be restricted (without decreasing the attainable sets), to take values such that $g(t, u) \in \partial g(t, U)$ (boundary controls), and even more, they can be restricted so that $g(t, u)$ is always an "extremal point" of the convex hull of g(t, U) .

Denote by ξ any orthogonal and normal basis of the space R^n, that is

$$\xi = (x_1, \ldots, x_n),$$

x_i (i=1, 2, ..., n) being n orthogonal vectors of length $|x_i| = 1$. Let (capital ξ) Ξ denote the set of all such bases :

$$\Xi = \left\{ \xi = (x_1, \ldots, x_n);\ x_i \in R^n,\ (x_i, x_j) = \delta_{ij} \right\} .$$

(Here $(x_i . x_j)$ is scalar product, δ_{ij} = 1 or = 0 according as to i = j or i ≠ j .)

Let A be an arbitrary compact subset of R^n. With each such A and each basis $\xi \in \Xi$ we associate a sequence A_j (j = 1, 2, ..., n) of subsets of A, defining :

(V.2.5) $$A_1 \left\{ x;\ x \in A,\ (x_1 . x) = \max_{y \in A} (x_1 . y) \right\} ,$$

(V.2.6) $$A_{j+1} = \left\{ x;\ x \in A_j,\ (x_{j+1} . x) = \max_{y \in A_j} (x_{j+1} . y) \right\} .$$

In words : A_1 maximizes the first coordinate in A, A_2 maximizes the second coordinate in A_1, and so on . Therefore A_1 is of dimension at most n-1, A_2 is of dimension at most n-2 , . . . , A_n is always a single point. This point is called the "lexicographic maximum" of the set A, and denoted by e(A , ξ)

For each compact A and $\xi \in \Xi$, the lexicographic maximum e(A, ξ) is an extremal point of A if A is convex, and of the convex hull of A if A itself is not convex. Conversely: to each extremal point $\underline{a}$ either of A if A is convex, or of the convex hull of A , there is some $\xi \in \Xi$ such that $\underline{a}$ = e(A, ξ). Let us recall that a point $\underline{a}$ of a convex set A is an extremal point of A if it is not an interior point of any interval contained in A (see Olech, [1]).

In order to study the control system (V.2.1) it is useful to make a coordinate transformation. Let $X(t)$ be the fundamental matrix solution of the homogeneous equation

(V.2.7) $$x' = A(t)\ x\ ,$$

so that $X'(t) = A(t) . X(t)$ and $X(0) = I$ = the unit matrix. The coordinate transformation

(V.2.8) $$x = X(t)\ y$$

straightens out the solutions of equation (V.2.7) , and the system (V.1.1) becomes

(V.2.9) $$y' = X^{-1}(t)\ g(t, u) = g_1(t, u)\ .$$

This system satisfies the same conditions assumed on (V.2.1) , and its solutions are simply

(V.2.10) $$y(t) = \int_{t_o}^{t} g_1(\theta , u(\theta))\ d\ \theta\ .$$

Let

(V.2.11) $$G(t) = \left\{ x;\ x = g_1(t, u) = X^{-1}(t)\ g(t, u)\ ,\ u \in U \right\} .$$

An admissible control $u(t)$ will be called <u>"extremal"</u> in the interval $I : t_o \leqslant t \leqslant t_1$, if there is a basis $\xi \in \Xi$ such that

(V.2.12) $$g_1(t, u(t)) = X^{-1}(t)\ \ g(t, u(t))\ = e(G(t) , \xi) \quad \text{a.e. in } I.$$

In recent papers, Olech ([1], [2], [3]) showed that the set

(V.2.13) $$E(t_1, t_o, x_o) = \left\{ x = x_u(t_1);\ x_u(t_o) = x_o\ ,\ u(t)\ \text{extremal} \right\} ,$$

(x_u = solution of (V.2.1) corresponding to the control $u(t)$), has the following properties:

i) $E(t_1, t_o, x_o)$ is the set of all extremal points of some compact and convex set $B(t_1, t_o, x_o)$, that is, there is some compact and convex $B(t_1, t_o, x_o)$ such that any point of $E(t_1, t_o, x_o)$ is an extremal point of $B(t_1, t_o, x_o)$ and any extremal point of $B(t_1, t_o, x_o)$ belongs to $E(t_1, t_o, x_o)$.

ii) The so defined set $B(t_1, t_o, x_o)$ coincides with the attainable set $F(t_1, t_o, x_o)$.

iii) If $x_1 \in F(t_1, t_o, x_o)$, $t_1 > t_o$, then there is an admissible control $u(t)$ and a sequence $\tau_o = t_o < \tau_1 < \tau_2 \ldots < \tau_k = t_1$, such that $u(t)$ is an extremal control in each of the intervals $[\tau_i, \tau_{i+i}]$ $(i = 1, 2, \ldots, k-1)$, and the integer k is not greater than $n+1$.

iv) Suppose $x_o(t)$ is a solution of (V.1.1) defined for all t. Then $x_o(t)$ is an extremal solution of (V.1.1) if and only if for each $t_1 < t_2$ and any other solution $x(t)$ of (V.1.1), the condition $x(t_i) = x_o(t_i)$ for $i = 1, 2$, implies that $x(t) = x_o(t)$ for $t_1 \leq t \leq t_2$.

These results have a very important intuitive meaning;

i) the attainable set is always compact and convex:

ii) every point (t, x) which can be reached with and admissible control, can also be reached with a piecewise extremal control, and the number of "switchings" from one extreme control to another is at most equal to n = dimension of the space;

iii) extreme controls are characterized by the fact that they lead to unique motions joining any two of their points (the corresponding control $u(t)$ may not be unique, because for a given value of $g(t, u)$ several possible values of u may be chosen).

For the proofs, the reader is referred to the papers mentioned above.

These results generalize the so called "bang-bang" principle, obtained by Bushaw ([1]) in a special case, by Bellman Glicksberg and Gross ([1]) for constant systems, and clearly formulated by LaSalle ([1]) for variable linear systems with U = a hypercube.

Extensions of LaSalle's bang-bang principle were given by Neustadt ([1]), Halkin ([3]) and Levinson ([1]). They are closely related with Lyapounov's theorem about the range of a vector measure (Lyapounov [1], for a shorter proof see also Halmos [1]) and its extension by Blackwell ([1], see also Halkin [4] [5], [1] and Castaing [1]).

Finally, it is worth mentioning a paper of Hermes ([2]), in which it is shown that any linear control problem like (V.1.1), where the set $g(t, U)$ is convex, can be approximated by another one where the new set $g_1(t, U)$ is strictly convex, i.e. all its boundary points are extremal points with positive gaussian curvature. Hence for the approximate problem, the optimal control given by Pontryagin's principle (which is, of course, an extremal control) is uniquely defined and continuous. This is not, in general, the case, when the set $g(t, U)$ is, for example, a polyhedron, because it is not strictly convex. By this method we are able to approximate an optimal control problem by another with continuous optimal control, uniquely defined by Pontryagin's principle, which might be an advantage, for example for computing purposes.

E.O. Roxin

VI. Controllability of linear systems.

§ 1. Some examples. In order to observe different types of characteristic behaviour od control systems, we start with some examples .

Example 1. Let the control equation be

$$x^{i} = -\lambda x^{i} + u^{i} \qquad (\lambda > 0,\ i=1,\ 2) \tag{VI. 1. 1}$$

with the restrictive condition on the control

$$(u^1)^2 + (u^2)^2 \leqslant k \quad . \tag{VI. 1. 2}$$

Starting from the origin $x_o^1 = x_o^2 = 0$, it is easy to see that the attainable set for $t \geqslant 0$ is (see figure VI. 1)

$$F(t, 0, 0) = \left\{ (x^1, x^2) \ ; \ \sqrt{(x^1)^2 + (x^2)^2} \leqslant \frac{k}{\lambda}(1-e^{-\lambda t}) \right\} .$$

For $t \to +\infty$, the attainable set tends to the disc of radius k/λ. But we can say more : any two points x_1, x_2 belonging to the open disc $(x_i^1)^2 + (x_i^2)^2 < (k/\lambda)^2$, are attainable from each other in a finite time. Indeed, we can even prescribe the curve $x = x(\theta)$ (θ is a parameter) along which we want to reach x_2 from x_1. For this purpose we have to determine $\theta = \theta(t)$ so that

$$x' = \frac{dx}{d\theta} \cdot \theta' = -\lambda x + u \ ,$$

hence,

$$u = \lambda x + \frac{dx}{d\theta} \quad \theta' \ ,$$

which is always possible, because as $|\lambda x|^2 < k$, there is some value $\theta' < 0$ such that $|\lambda x + \frac{dx}{d\theta} \quad \theta'|^2 < k$.

If the initial condition for (VI. 1. 1-2) is $x(0) = x_o$, the attainable set $F(t, 0, x_o)$ is obtained easily; it is represented in figure VI. 2 .

Example 2. Let

$$x^{i\prime} = \lambda x^i + u^i \qquad (\lambda > 0,\ i = 1, 2), \tag{VI. 1. 3}$$

with the same conditinns(VI. 1. 2) Now, for $x(0) = 0$,

$$F(t, 0, 0) = \left\{ x ; \ |x| \leqslant \frac{k}{\lambda}(e^{\lambda t} - 1) \right\} .$$

(See figure VI. 3) For $t \rightarrow +\infty$, the attainable set covers the whole plane. But it should be noticed that once a point outside the disc of radius k/λ is reached, one cannot return to the origin (or points nearby) . This results from considering the attainable set from any x_o, which is shown in figure VI. 4 .

In all these examples, as long as we stay inside the disc of radius k/λ , or, as is equivalent, as long as k in (VI. 1. 2) is large enough, every point is attainable from any other point, in finite time.

Example 3. Consider the same control equation (VI. 1. 1) with the condition

$$-1 \leqslant u^i \leqslant +1 \qquad (i=1, 2) . \tag{VI. 1. 4}$$

All the considerations to be made , are similar to example 1; figure VI. 5, similar to figure VI. 2.

Example 4. Let

$$x^{i\prime} = - \lambda x^i + k^i u \qquad (\lambda > 0 ,\ i=1, 2) , \tag{VI. 1. 5}$$

with

(VI. 1;6) $$-1 \leqslant u \leqslant +1 .$$

(Here the dimension of the control u is less than the dimension of the state variable x.)

The linear combination $y = k^1 x^2 - k^2 x^1$ satisfies the differential equation $y' = -\lambda y$. The boundary controls $u = \pm 1$ give the solutions

$$x^i(t) = x_o^i e^{-\lambda t} \pm \frac{k^i}{\lambda} (1-e^{-\lambda t}) .$$

Therefore the attainable set is as shown in figure VI. 6. The new feature in this example is that the attainable set has dimension one.

<u>Example 5.</u> Let

(VI. 1. 7) $$x^{i\,\prime} = -\lambda_i x^i + k^i u \qquad (\lambda_i > 0,\ i = 1 , 2) ,$$

with the same condition (VI. 1. 6) . If the characteristic roots λ_1, λ_2 are equal, we have the preceding example. If they are distinct, however, the result is quite different.

To discuss the solutions of (VI. 1. 7) , it is sufficient to consider the boundary controls $u = \pm 1$. Equation (VI. 1. 7) can then be written

$$(x^i \mp \frac{k^i}{\lambda_i})' = -\lambda_i (x^i \mp \frac{k^i}{\lambda_i}) ,$$

so that the trajectories are parabolas converging to the points $x^i = \pm \frac{k^i}{\lambda_i}$ (figure VI. 7) .Combining these trajectories, it is seen that in the leaf-shaped region, each point is attainable from each other.

More examples of this type are given in the paper of Roxin-Spinadel ([1]) (Roxin ([1])). From the examples given here, it is already clear that the dimension of the attainable set is an important feature. It is also observable that the boundedness condition on the control ($u \in U$) plays little role in the determination of this dimension. Therefore it is reasonable to start considering the case with unbounded control, $u \in R^m$.

§ 2. <u>Controllability and observability.</u> These concepts were introduced by Kalman ([1], [2], [3], [4]) Kalman, Ho and Narendra ([1]) for linear systems. Here we understand as linear system, a physical system describable by equations:

$$\text{(VI. 2. 1)} \qquad \begin{cases} x' = A x + B u \\ v = C x + D u \end{cases}$$

where x = n-vector is the "state" of the system, u = m-vector is the "input" (control), v = r-vector is the "output", A, B, C, D are matrices which are, in general, functions of time. The only way to influence, from the outside, the evolution of the system, is through the input $u(t)$, the only way to detect, from the outside, the behaviour of the system, is through the output $v(t)$. These relations are indicated graphically in figure VI. 8. (Kalman really states the problem in a more abstract way, but for our purposes it is sufficient to consider it as indicated here).

In order to simplify things, assume that the matrices A, B, C, D are constant and that the characteristic roots of A, λ_i, are distinct. Then, after a coordinate transformation, we may suppose that

A is in diagonal form. In the "free" system (with input $u = 0$) there is, then, no mutual influence between the different components $x^i(t)$. In this case, if there are some rows of B made up by zeros, these correspond to components x^i which are not influenced by the control. These components are called "uncontrollable", the others, "controllable". Similarly, the columns of C which are zero, corresponds to components x^i which do not influence the output v: they are "unobservable". The other components are called "observable".

By ordering the components x^i in such a way that components of similar behaviour are together, the system (VI. 2. 1) can be decomposed into 4 subsystems :

1) a system S_1 which is "completely controllable and observable",
2) a system S_2 which is "controllable" but "unobservable",
3) a system S_3 which is "observable" but "uncontrollable",
4) a system S_4 which is "uncontrollable" and "unobservable".

Schematically, this is indicated in figure VI. 9. This constitutes, essentially, the "canonical decomposition theorem" of Kalman ([3]).

For time-variable matrices in the general case, the relations are less clear, and different definitions of controllable and observable systems can be given, which coincide for the above mentioned case of constant matrices with distinct roots. Here we prefer the following definition.

Definition. The system (VI. 2. 1) is called "completely controllable at time t_o", if any state x_1 can be attained in finite time $t_1 > t_o$ from any other state x_o at time t_o.

Note that a system which is controllable at t_o, is also controllable at every $t < t_o$ but not necessarily at $t > t_o$.

In the definition of controllability, the second of the equations (VI. 2. 1) does not intervene; therefore we do not need to consider it when dealing with the problem of controllability.

There is a great similarity between the concepts of controllability and observability. These concepts are "dual", in the sense that for every property of controllability, a corresponding one for observability can be given.

§ 3. Conditions for complete controllability. Here we will follow a paper of Hermes ([1]). The equation under consideration will be.

(VI. 3. 1) $$x' = A(t)\ x + B(t)\ u \qquad (x = n\text{-vector},\ u = m\text{-vector}),$$

where A(t) is L_1 and B(t) is L_2 in every finite interval. Admissible controls u(t) are all measurable functions $u : R \to R^m$.

Lemma. A necessary and sufficient condition that there exists an $m \times n$ matrix valued function V(t) in $L_2\ [t_o, t_1]$ such that for some $t_1 > t_o$,

(VI. 3. 2) $$\int_{t_o}^{t_1} B(\tau)\ V(\tau)\ d\tau$$

is non-singular, is that for some $t_1 > t_o$,

(VI. 3. 3) $$\int_{t_o}^{t_1} B(\tau)\, B^T(\tau)\, d\tau$$

is non-singular. (B^T = traspose of B).

Proof : sufficiency is immediate choosing $V(\tau) = B^T(\tau)$. To show necessity, assume there exist V, $t_1 > t_o$, such that (VI. 3. 2) is non-singular, but

$$\int_{t_o}^{t} B(\tau)\ B^T(\tau)\, d\tau$$

is singular for all $t > t_o$. This implies there exists a constant vector $c \neq 0$ such that

$$c\left[\int_{t_o}^{t_1} B(\tau)\, B^T(\tau)\, d\tau\right] c^T = 0 \ ,$$

and since $B(\tau)\, B^T(\tau)$ is positive semidefinite, we obtain $c\, B(t) = 0$ almost everywhere in $[t_o, t_1]$. Thus

$$\int_{t_o}^{t_1} cB(\tau)\ V(\tau)\, d\tau = 0 \ ,$$

contradicting the nonsingularity of $\int_{t_o}^{t_1} B(\tau)\ V(\tau)\ d\tau$.

Theorem I. A necessary and sufficient condition for the system

(VI. 3. 4) $\qquad x' = B(t)\ u$

to be completely controllable at t_o, is that there exists $t_1 > t_o$ such that

(VI. 3. 5) $\qquad M(t_o, t_1) = \int_{t_o}^{t_1} B(\tau)\, B^T(\tau)\ d\tau$

is non-singular.

Sufficiency: Let x_o, x_1 be any two points in R^n. We will show that (t_1, x_1) is attainable from (t_o, x_o). We may take $u(t) = B^T(t)\,\xi$, where $\xi = \text{const} \in R^n$. We desire

$$x_1 = x_o + \left[\int_{t_o}^{t_1} B(\tau)\, B^T(\tau)\, d\tau\right] \xi \ ,$$

or $\xi = M^{-1}(t_o, t_1)(x_1 - x_o)$.

Necessity : Assume $M(t_o, t_1)$ is singular for every $t_1 > t_o$. This implies, as in the proof of the preceding lemma, that there is a constant vector $c \neq 0$ such that $c\,B(t) \equiv 0$ a.e. Since x_o is arbitrary, let it be such that $c\,x_o = 0$. We will show that the point c is not attainable from x_o. Suppose, indeed, for some u and t_1, $c = x_o + \int_{t_o}^{t_1} B(\tau)\, u(\tau)\, d\tau$. Then

$$c.c. = |c|^2 = c.x_o + c.\int_{t_o}^{t_1} B(\tau)\, u(\tau)\, d\tau = 0 ,$$

a contradiction to the fact that $c \neq 0$.

<u>Theorem II.</u> (Kalman). The linear system (VI. 3. 1) is completely controllable at t_o, if and only if

$$\text{(VI. 3. 6)} \qquad \int_{t_o}^{t_1} X(t_o, \tau)\, B(\tau)\, B^T(\tau)\, X^T(t_o, \tau)\, d\tau$$

is nonsingular for some $t_1 > t_o$. Here $X(t, \tau)$ is a fundamental solution of the homogeneous system $x' = A(t)\, x$.

Proof: the transformation

$$y(t) = X^{-1}(t, t_o)\, x(t) = X(t_o, t)\, x(t)$$

transforms the differential equation (VI. 3. 1) in

$$\text{(VI. 3. 7)} \qquad y'(t) = X(t_o, t)\, B(t)\, u(t) .$$

The transformation being nonsingular, the system (VI. 3. 1) is completely controllable if and only if the system (VI. 3. 7) is completely controllable, i.e. according to theorem I, if (VI. 3. 6) is nonsingular for some $t_1 > t_o$.

<u>Theorem III</u>(Kalman). For constant systems (matrices A and B in (VI, 3.1) constant), complete controllability is equivalent with the condition

(VI. 3. 8) $$\text{rank}\,[B,\ A\cdot B,\ A^2\cdot B,\ \ldots,\ A^{n-1}\cdot B] = n .$$

It has to be shown that for $X(t) = e^{At}$, (VI. 3. 8) is equivalent to (VI. 3. 6) nonsingular.

Assuming (VI. 3. 6) to be singular, there exists a vector $c \neq 0$ such that $c.e^{At}.B \equiv 0$ (we take $t_o = 0$, $t_1 = t$). Differentiating, we obtain successively

$$c.e^{At}\cdot A\cdot B = 0,\quad c.e^{At}.A^2.B = 0, \ldots,\ c.e^{At}.A^{n-1}.B = 0.$$

As $c.e^{At} \neq 0$,

(VI. 3. 9) $$\text{rank}\,[B,\ A.B,\ A^2.B, \ldots,\ A^{n-1}.B] < n$$

follows.

Assuming (VI. 3. 9), there is a vector $c \neq 0$ which is orthogonal to every column of the matrix of (VI. 3. 9), hence

$$c.B = 0,\ c.A.B = 0, \ldots,\ c.A^{n-1}.B = 0 .$$

Therefore, also,

$$c.A^n.B = 0,\ c.A^{n+1}.B = 0,\ \ldots .$$

Indeed, by the Cayley-Hamilton theorem, the matrix A satisfies its own characteristic equation, and therefore A^n is expressable as a linear combination of $I, A, A^2, \ldots, A^{n-1}$, and the same happens to all higher powers of A.

But then also

$$X(t) = e^{At} = I + \frac{t}{1} A + \frac{t^2}{2} A^2 + \ldots$$

satisfies

$$c.X(t).B = 0 ,$$

and the singularity of (VI. 3. 6) follows.

Corollary . The n-th order control equation with constant coefficients

(VI. 3. 10) $$x^{(n)} + a_{n-1} x^{(n-1)} + \dots + a_1 x' + a_o x = u ,$$

(u = scalar = control function) is completely controllable.

It is, indeed, a simple matter of computation to see that the condition of theorem III is satisfied with

$$A = \begin{pmatrix} 0 & 1 & 0 & \dots & 0 \\ 0 & 0 & 1 & \dots & 0 \\ . & . & . & . & . \\ 0 & 0 & 0 & & 1 \\ -a_o & -a_1 & -a_2 & \dots & -a_{n-1} \end{pmatrix} \qquad B = \begin{pmatrix} 0 \\ 0 \\ \vdots \\ 0 \\ 1 \end{pmatrix}$$

Hermes ([2]) extended this result to the n-th order control equation (VI. 3. 10) with variable coefficients $a_{n-1}(t)$, $a_{n-2}(t), \dots$ $a_o(t)$, showing that it satisfies the condition of theorem II.

§ 4. Controllability under boundedness conditions. In this chapter we have developed the concept of controllability, disregarding boundedness restrictions on the control. If we take these into account, i. e. if we consider actual attainability, there are also interesting results.

Here we want to mention only that, in a very general way, boundedness conditions can be regarded as boundedness in some norm; $\|u\| \leqslant 1$. This includes the most common condition $|u^i(t)| \leqslant 1$ for every t, but also cases like $\|u\| = \|u\|^p_r$, where the r-norm of the vector u^i is taken as

$$|u|_r = \left[\sum (u^i)^r \right]^{1/r}$$

and the p-norm of the function $\mu(t)$, as

$$\|u\|_r^p = \left[\int (|u(t)|_r)^p \, dt \right]^{1/p}$$

This general theory was developed recently by Antosiewicz ([1]) and Conti ([1]) , and we refer the reader to these papers. It should be noted also that with these general kind of boundedness conditions, optimal controls may appear which are "impulse"-functions (of the type of the "delta" of Dirac)(Rishel, [1], Schmaedecke, [1]) . The formulation of the control problem and many important results can be made , without major changes, for linear systems in Banach spaces (Fattorini [1],[2] , Balakrishnan [1]) . The importance of this lies in the possible application to much more general systems, for example partial differential equations.

The relation with problem of moments was pointed out, among others, by Neustadt ([2], [3]) .

E. O. Roxin

VII. Controllability of nonlinear systems.

§1. Attainability and controllability. For the general nonlinear control problem

(VII. 1. 1) $x' = f(t, x, u)$ (x = n-vector, u = m-vector),

the concept of controllability is by far not as clear as for linear systems. First at all, there is no essential difference between the case of bounded and unbounded control ($u \in U$, U compact and $U = R^m$ respectively). To see this it is sufficient to consider possible control equations of the type $x' = f(t, x, \sin u)$, $U = R$, which is practically a bounded control, and $x' = f(t, x, \tan u)$, $|u| < \frac{\pi}{2}$, where the control is practically unbounded.

The problem of optimal control has been formulated very often (see, for example, Bellman et al. [1] or LaSalle [2]) as the problem of finding a control which steers the system, from an initial state x_o at $t = t_o$, to the origin $x_1 = 0$ in some optimal way. In connection with this, a state x_o (at $t = t_o$) is called controllable if there exists an admissible control which steers the system from (t_o, x_o) to $(t_1, 0)$ for some finite $t_1 > t_o$. The set of points (t_o, x_o) which are controllable in this sense, is then called set of controllability. Obviously, changing the time direction t into $-t$, this set is the set of attainability from the origin (at the corresponding time).

It would be quite natural to say that system (VII. 1. 1) is completely controllable, if the corresponding set of controllability is the whole R^{n+1}. Newertheless, with this definition, some patological behaviour of the system may happen which is better excluded. Therefore we will give later the definition of controllability according to Hermes. First we will see some criteria of attainability.

Theorem I (Hermes) : Consider the control equation

(VII. 1. 2) $x'(t) = g(t, x(t)) + B(t)\,u(t)$

where x, g = n-vectors, u = m-vector , B = n × m-matrix, satisfying the following assumptions :

i) g is continuous as a function of t, for each x.

ii) $|g^j(t, x)| \leqslant N$, $j = 1, 2, \dots, n$.

iii) $|g^j(t, x) - g^j(t, \bar{x})| \leqslant m.\,|x-\bar{x}|$, $j = 1, 2, \dots, n$.

iv) $B(t)$ is L_2 in any finite t-interval

Then a sufficient condition that all points $x \in R^n$ be attainable from a given (t_o, x_o), for some $t_1 \geqslant t_o$, with controls $u(t) \in L_2(R \to R^m)$, is that

(VII. 1. 3) $$M(t_o, t_1) = \int_{t_o}^{t_1} B(\tau)\, B^T(\tau)\, d\tau$$

be nonsingular.

Denoting by φ^u the solution of (VII. 1. 2) with $\varphi^u(t_o) = x_o$ corresponding to the control u,

(VII. 1. 4) $$\varphi^u(t) = x_o + \int_{t_o}^{t} g(\tau, \varphi^u(\tau))\, d\tau + \int_{t_o}^{t} B(\tau)u(\tau)\, d\tau .$$

Let $\bar{x} \in R^n$ be an arbitrary point. To find a control $u(t)$ such that $\varphi^u(t_1) = \bar{x}$, it is sufficient to consider the n-parameter family of controls $u(t) = B^T(t)\,.\,\xi$, where $\xi \in R^n$

The corresponding solutions will be denoted by $\varphi^{\xi}(t)$.

Define the mapping $\mathcal{F}: R^n \to R^n$ as follows.

Let

$$\alpha(\xi) = \int_{t_o}^{t_1} g(\tau, \varphi^{\xi}(\tau))\, d\tau ,$$

and define $\mathcal{F}(\xi) = M^{-1}(t_o, t_1)[\bar{x} - \alpha(\xi) - x_o]$. From (VII. 1. 4) it follows that a fixed point of $\mathcal{F}$ will yield a value ξ such that $\varphi^{\xi}(t_1) = \bar{x}$

The conditions imposed on g(t, x) assure that φ^{ξ} is a continuous function of ξ in the topology $C[t_o, t_1]$, i. e. the topology induced by the supremum norm . Thus $\alpha(\xi)$ is a continuous function of ξ, and $\mathcal{F}$ is a continuous function of ξ.

We next show that there is a K such that $|\xi| \leq K$ implies $|\mathcal{F}(\xi)| \leq K$. Letting $|\xi| = \sum |\xi_i|$ and $|M^{-1}|$ be any matrix norm, since $|g^j| \leq N$, for any ξ, $|\alpha(\xi)| \leq n(t_1 - t_o) N$. Letting

$$K = |M^{-1}(t_o, t_1)| [|\bar{x}| + n N(t_1 - t_o) + |x_o|] ,$$

it follows that for any ξ, $|\mathcal{F}(\xi)| \leq K$; hence, in particular, $\mathcal{F}$ maps the ball $\{\xi \in R^n ; |\xi| \leq K\}$ continuously into itself . Thus $\mathcal{F}$ has a fixed point.

In this theorem, the conditions imposed on g(t, x) exclude linear systems. These are inclued in the following stronger version of the same theorem.

<u>Theorem II</u>. Let the control equation be

(VII. 1. 5) $\qquad x' = A(t)\, x + g(t, x) + B(t)\, u ,$

where A(t) is an integrable n×n-matrix, and all the assumptions of theorem I are fulfilled. Let $X(t, t_o)$ be a fundamental solution of the equation $x' = A(t)\, x$. Then , if

(VII. 1. 6) $\qquad \int_{t_o}^{t_1} X(t_o, \tau) B(\tau)\, B^T(\tau)\, X^T(t_o, \tau)\, d\tau$

is nonsingular, $t_1 > t_o$, then all $\bar{x} \in R^n$ are attainable from (t_o, x_o) (x_o arbitrary) fro $t = t_1$, with L_2 controls.

Make the coordinate transformation

$$x = X(t, t_o) \, . \, y \, .$$

Then $x' = A.X.y + X.y'$ and y satisfies the equation

(VII.1.7) $$y' = X^{-1}(t, t_o) \, . \, g(t, X(t, t_o) \, . \, y) + X^{-1}(t, t_o) . B(t) \, . \, u \, .$$

In the interval $[t_o, t_1]$, $X(t, t_o)$ is continuous, bounded and nonsingular, the same as $X^{-1}(t, t_o) = X(t_o, t)$. Therefore (VII.1.7) satisfies the conditions of theorem I, and the result follows.

§ 2. Nonlinear systems with linear control. Definition of complete controllability. We consider control equations of the type

(VII.2.1) $$x' = g(t, x) + B(t, x) \, u \, ,$$

where x, g = n-vectors, u = m-vector, B = n×m-matrix. This is not really a restriction, because in the general case $x' = f(t, x, u)$ we could always write $x' = f(t, x, 0) + v$, with $v = f(t, x, u) - f(t, x, 0)$, so that the control equation is in the form (VII.2.1), with B = the identity matrix. The restriction consists in considering, as we will do, unbounded controls $u(t)$ with the only restriction of being L_2.

It will be assumed that g and B are C^1 functions in all arguments. Also the assumptions $1 \leq m < n$ and rank $B = m$ are made. This excludes trivial cases ($m = 0$ is a system without control, rank $B < m$ means that there are some "ineffectual" control components, and $m = n$ has all possible functions $x(t)$ as solutions)

Take, at each point (t, x), a $(n-m) \times n$-matrix $H(t, x)$ of maximum rank, such that

(VII. 2. 2) $$H(t, x) \cdot B(t, x) \equiv 0 .$$

This is possible because there are exactly $n - m$ linearly independent n-vectors (the rows of H), orthogonal to the m columns of B.

Consider the pfaffian system

(VII. 2. 3) $$H(t, x)\, dx - H(t, x)\, g(t, x)\, dt = 0 .$$

(Note that this systems consists of $n-m$ scalar equations.)

Every solution $x(t)$ of (VII. 2. 1) satisfies

$$dx = g(t, x)\, dt + B(t, x)\, u(t)\, dt .$$

Multiplying by $H(t, x)$, and taking into account (VII. 2. 2),

(VII. 2. 4) $$H(t, x)\, dx = H(t, x)\, g(t, x)\, dt,$$

so that $x(t)$ satisfies (VII. 2. 2).

If $c = c(t, x)$ is an n-m row-vector , then

(VII. 2. 5) $$h(t, x) = c(t, x) \cdot H(t, x)$$

is an n-vector which is a linear combination (at each point (t, x)) of the n-m rows of H. Then, by (VII. 2. 4),

(VII. 2. 6) $$h(t, x)\, dx - h(t, x)\, g(t, x)\, dt = 0$$

along each solution $x(t)$ of the control equation (VII. 2. 1).

The pfaffian system (VII. 2. 3) is said to be <u>integrable</u> at (t_o, x_o), if there exists a C^1 scalar valued function $\psi(t, x)$ and an $\varepsilon > 0$, such that

(VII. 2. 7) $\psi_x(t, x) = h(t, x)$, $\psi_t(t, x) = - h(t, x)\, g(t, x)$

for $|t - t_o| < \varepsilon$, $|x - x_o| < \varepsilon$, for some $h(t, x)$ as in (VII. 2. 5).

According to this definition, if the pfaffian is integrable, every solution $x(t)$ of the control equation (VII. 2. 1) remains on the hypersurface $\psi(t, x) = \text{const.}$, as long as $x(t)$ remains in the ε-neighborhood of (t_o, x_o). Indeed, along $x(t)$,

$$\frac{d}{dt}\,\psi(t, x(t)) = \psi_t(t, x(t)) + \psi_x(t, x(t))\, x'(t) =$$

$$= \psi_t(t, x(t)) + \psi_x(t, x(t))\,(g(t, x(t)) + B(t, x(t))u(t)) = 0 ,$$

by virtue of (VII. 2. 7).

Note that in the case of an integrable pfaffian, points (t, x) which do not belong to the surface $\psi(t, x) = \text{const.}$ may be attainable from (t_o, x_o) by solutions $x(t)$ of the control equation, which go outside the mentioned ε-neighborhood.

According to Hermes ([2]) we will give the following definition.

The control system (VII. 2. 1) is <u>completely controllable</u> at (t_o, x_o) if the associated pfaffian system (VII. 2. 3) is not integrable at (t_o, x_o).

It has to be shown that for linear systems, this definition coincides with condition (VI. 3. 6) of theorem II of chapter VI.

For the linear system

(VII. 2. 8) $x' = A(t)\, x + B(t)\, u$,

in the corresponding pfaffian equation (VII. 2. 6) $h(t, x)$ may be taken as a function of t alone

(VII. 2. 9) $h(t)\, dx - h(t)\, A(t)\, x\, dt = 0$

($h(t)$ is here any vector orthogonal to all columns of $B(t)$).

Indeed, if (VII. 2. 9) has an integrating factor $\mu(t,x)$, there is another independent of x. To see this, assume

$$\mu(t,x)\, h(t)\, dx - \mu(t,x)\, h(t,x)\, A(t)\, x\, dt$$

is an exact differential. Then

$$\mu_{x_j} h_i - \mu_{x_i} h_j = 0 \quad (i, j = 1, 2, \dots, n)$$

and

$$\mu_t\, b + \mu b' = -\mu_x\, b\, A\, x - \mu b\, A .$$

Therefore $\mu_o(t) = \mu(t, 0)$ is also an integrating factor.

Therefore the pfaffian system associated with (VII. 2. 8) may be taken as

(VII. 2. 10) $$H(t)\, dx - H(t)\, A(t)\, x\, dt = 0 .$$

According to the last definition, (VII. 2. 8) will be completely controllable at (t_o, x_o) if and only if (VII. 2 . 10) is not integrable at (t_o, x_o), i. e. if there is no $\psi(t,x)$ which, in an ε-neighborhood of (t_o, x_o), satisfies

(VII. 2. 11) $$\psi_x(t,x) = h(t) , \qquad \psi_t(t,x) = -h(t)\, A(t)\, x .$$

Let $X(t, t_o)$ be a fundamental solution of $x' = A(t)\,x$. Define

(VII. 2. 12) $$W(t_o, t_1) = \int_{t_o}^{t_1} X(t_o, t)\, B(t)\, B^T(t)\, X^T(t_o, t)\, dt.$$

According to chapter VI, $W(t_o, t_1)$ nonsingular $(t_1 > t_o)$ is equivalent with the condition that (t_1, x_1) is attainable from (t_o, x_o) for every x_o, x_1. If we want a property of local character, we have to demand that t_1 could be taken as near t_o as we want. But then $W(t_o, t_1)$ must be nonsingular for all $t_1 > t_o$, because $W(t_o, t_1)$ is positive semidefinite, hence $W(t_o, t_1)$ nonsingular implies $W(t_o, t_2)$ nonsingular for every

$t_2 > t_1$.

Theorem I (Hermes). A necessary and sufficient condition that $W(t_o, t_1)$ be nonsingular for all $t_1 > t_o$ is that the pfaffian (VII. 2. 10) be not integrable at t_o.

To prove necessity, we show that if the pfaffian (VII. 2. 10) is integrable at t_o, then $W(t_o, t_1)$ is singular for some $t_1 > t_o$.

Assume the pfaffian (VII. 2. 10) integrable at t_o. Then there is a vector, h, orthogonal to B, such that in $t_o \leqslant t < t_o + \varepsilon$ for some $\varepsilon > 0$, (VII. 2. 11) is satisfied and therefore $h' = - h A$, which is the adjoint equation of $x' = A x$. Then $h(t) = c X^{-1}(t, t_o) =$

$= c X (t_o, t)$ for some constant vector c. As $h(t). B(t) = 0$,

$$c. X(t_o, t). B(t) = 0$$

in $t_o \leqslant t < t_o + \varepsilon$ and $W(t_o, t_1)$ in (VII, 2, 12) is singular for some $t_1 > t_o$.

To prove sufficiency, we show that if $W(t_o, t_1)$ is singular for some $t_1 > t_o$, then (VII. 2. 10) is integrable at $t = t_o$.

Assume $W(t_o, t_1)$ is singular, $t_1 > t_o$; Then $W(t_o, t)$ is singular for all $t_o \leqslant t \leqslant t_1$. This implies there exists a vector $c(t_1)$ such that $c(t_1). W(t_o, t_1) c^T(t_1) = 0$. Since the integrand in (VII. 2. 12) is continuous and positive semidefinite,

$$c(t_1). X(t_o, t). B(t). B^T(t). X^T(t_o, t). c^T(t_1) = 0$$

$$\text{for } t_o \leqslant t \leqslant t_1 .$$

It follows that $0 \equiv c(t_1). X(t_o, t). B(t) = c(t_1). X^{-1}(t, t_o) . B(t)$. Thus h defined by $h(t) = c(t_1). X^{-1}(t, t_o)$ is an admissible vector, in the sense that $h(t). B(t) \equiv 0$.

Define the scalar valued function $\psi(t, x) = c(t_1). X^{-1}(t, t_o) . x$.

Then $\psi_x(t,x) = h(t)$, $\psi_t(t,x) = -h(t).A(t).x$, showing that the pfaffian (VII. 2.10) is integrable.

We proceed now to examine the condition of controllability along a certain solution. Let the initial condition be $x(t_o) = 0$, $v \in L_2$ an arbitrary control and φ^v the corresponding solution of (VII. 2.1) . Let $u(t;\xi)$ be an n-parameter family of controls, $\xi \in R^n$, such that $u(t;0) = v(t)$, $u_\xi = \frac{\partial u}{\partial \xi}$ exists and denote by $x(.\,;\xi)$ the solution corresponding to $u(.\,;\xi)$. Then $x(.\,;\xi)$ satisfies

$$x(t;\xi) = \int_{t_o}^{t} [g(\tau, x(\tau;\xi)) + B(\tau, x(\tau;\xi))\, u(\tau;\xi)]\, d\tau ,$$

$$x_\xi(t;0) = \int_{t_o}^{t} \{ [g_x(\tau,\varphi^v(\tau)) + B_x(\tau,\varphi^v(\tau))\, v(\tau)]\, x_\xi(\tau;0) + B(\tau,\varphi^v(\tau))\, u_\xi(\tau;0) \}\, d\tau ,$$

where $B_x.v$ is the matrix with i, j-th element

$$\sum_{\nu=1}^{r} B^{i\nu}_{x_j} v^\nu \quad (i,j=1,2,\ldots,n) .$$

For each $t \geq t_o$, $x(t;\xi)$ is a mapping $\xi \to x$ with $0 \to \varphi^v(t)$. Let $Z(t;\varphi^v, u_\xi)$ denote the Jacobian matrix $x_\xi(t;0)$. If this Jacobian is, for a certain $\bar{t}$, nonsingular , then the implicit function theorem assures that all points x of a certain n-dimensional neighborhood of $\varphi^v(\bar{t})$, are attainable at $t = \bar{t}$ (from $(t_o, 0)$ and this using only controls of the family $u(t;\xi)$) .

Let $X(t,t_o)$ be a fundamental solution matrix of the system

$$x'(t) = [g_x(t,\varphi^v(t)) + B_x(t,\varphi^v(t))\, v(t)]\, x(t) .$$

Then

$$Z(t; \overset{v}{\varphi}, u_{\xi}) \equiv \int_{t_o}^{t} X(t,\tau)\, B(\tau, \overset{v}{\varphi}(\tau))\, u_{\xi}(\tau;0)\, d\tau .$$

Then , from the lemma and theorem II of chapter VI we have:

Theorem II (Kalman) . A necessary and sufficient condition that there exists an m×n-matrix u_{ξ} such that $Z(t_1; \overset{v}{\varphi}, u_{\xi})$ is nonsingular for some $t_1 > t_o$, is that the linear system

$$y'(t) = \left[g_x(t, \overset{v}{\varphi}(t)) + B_x(t, \overset{v}{\varphi}(t))\, v(t) \right] y(t) + B(t, \overset{v}{\varphi}(t)) u(t)$$

is completely controllable.

In terms of the Pfaffian approach the equivalent theorem is:

Theorem III (Hermes) . A necessary and sufficient condition that there exists an m×n-matrix u_{ξ} such that $Z(t_1; \overset{v}{\varphi}, u_{\xi})$ is nonsingular for some $t_1 > t_o$, is that the Pfaffian system

$$H(t, \overset{v}{\varphi}(t))\, dx - H(t, \overset{v}{\varphi}(t)) \left[g_x(t, \overset{v}{\varphi}(t)) + H_x(t, \overset{v}{\varphi}(t)) v(t) \right] x\, dt = 0$$

be nonintegrable for some $t_1 > t_o$.

§ 3. General nonlinear systems. Here the situation is even more complicated as in § 2, and we want to mention only a result similar to theorem II and III of § 2 .

The general equation is

(VII. 3. 1) $\quad x' = f(t, x, u) , \qquad x(t_o) = x_o ;$

As usually, x = n-vector, u = measurable, unrestricted m-vector valued control. f is supposed C^2. Given a certain control v(t) and the corresponding solution $\overset{v}{\varphi}(t)$, system (VII. 3. 1) is said to be locally controllable along the solution $\overset{v}{\varphi}(t)$, if for some $t_1 > t_o$,

E. O. Roxin

all points in some state space (n-dimensional) neighborhood of $\overset{v}{\varphi}(t_1)$ are attainable in time t_1 by solutions of (VII. 3. 1).

It should be noted that this definition requires more than just that all points in the mentioned neighborhood should be attainable (at some time possibly different from t_1). For example, $\overset{v}{\varphi}(t_1)$ is on the boundary of the attainable set $F(t_1, t_o, x_o)$, the requirement of the definition cannot be fulfilled (at least for that value of t_1).

Under the stated conditions ($f \in C^2$), it is reasonable to consider, along $\overset{v}{\varphi}(t)$, the linear approximation of (VII. 3. 1)

(VII. 3. 2) $$y'(t) = f_x(t, \overset{v}{\varphi}(t), v(t)) . y(t) + f_u(t, \overset{v}{\varphi}(t), v(t)). w(t)$$

where $y = x(t) - \overset{v}{\varphi}(t)$ and $w = u(t) - v(t)$.

This is a linear control equation of the type considered in § 2. The same method used to obtain, there, theorem III, gives here the following result.

<u>Theorem I.</u> (Hermes). A sufficient condition that there exists a $t_1 \geqslant t_o$, such that all points in some state space neighborhood of $\overset{v}{\varphi}(t_2)$ for all $t_2 > t_1$ are attainable in time t_2 by solutions of (VII. 3. 1), is that there exists a $t_1 \geqslant t_o$ such that the Pfaffian system

$$H(t;v)\, dy - H(t;v)\, f_x(t, \overset{v}{\varphi}(t), v(t)) = 0$$

is not integrable at t_1; here $H(t;v)$ is any $(n-m) \times n$-matrix of maximum rank such that $H(t;v) . f_u(t, \overset{v}{\varphi}(t), v(t)) \equiv 0$.

VIII. Singular problems. Not controllable systems.

§ 1. Systems linear in the control. Consider the control equation

(VIII. 1. 1) $x' = g(t, x) + B(t, x).u \; , \quad x(t_o) = x_o \; ,$

with the same assumptions of chapter VII (VII. 2. 1) . We have already seen that a sufficient condition for local controllability of a certain solution $\varphi^v(t)$ of (VIII. 1. 1) corresponding to the control v(t) , is that there exists an (n - m)×n-matrix u_ξ such that the Jacobian

(VIII. 1. 2) $$\left.\frac{\partial x(t, \xi)}{\partial \xi}\right|_{\xi=0} = Z(t, \varphi^v, u_\xi)$$

is nonsingular for some $t > t_o$. (Here, as in chapter VII, we assume an n-parameter family of controls $u(t; \xi)$, $\xi \in R^n$, such that, $u(t;0) = v(t)$ and $x(t, \xi)$ is the solution of (VIII. 1. 1) corresponding to $u(t) = u(t; \xi)$.) By theorem III (VII. 2) this condit is equivalent to the non integrability, for some $t > t_o$, of the Pfaffian system

(VIII. 1. 3) $H(t, \varphi^v(t))dx - H(t, \varphi^v(t))\left[g_x(t, \varphi^v(t)) + B_x(t, \varphi^v(t))v(t)\right] x \, dt = 0 \; .$

In other words, for some $t > t_o$ and any h(t, x) orthogonal to every column of B(t, x) .

(VIII. 1. 4) $h(t, \varphi^v(t)) \, dx - h(t, \varphi^v(t)) \left[g_x(t, \varphi^v(t)) + B_x(t, \varphi^v(t))v(t)\right] x \, dt$

is not an exact differential.

Let us see what happens if (VIII. 1. 4) is an exact differential. Then (and only then)

(VIII. 1. 5) $$\frac{d}{dt} h(t, \varphi^v(t)) = - h(t, \varphi^v(t)) \left[g_x(t, \varphi^v(t)) + B_x(t, \varphi^v(t))v(t)\right]$$

which is precisely the adjoint system of Pontryagin's maximum principle, determining the vector p (IV. 2. 7), for the time-optimal control problem. Therefore the vector h in (VIII. 1. 5) is a possible adjoint vector for this problem. According to the maximum principle, the control u(t) has to be chosen (in order to be optimal) so that

$$h(t).f(t,x,u) = h(t)\left[g(t,x) + B(t,x).u\right] = \text{maximum a.e.}$$

As $h(t).B(t, \varphi^v(t)) \equiv 0$, it is seen that in the integrable (not controllable) case, the maximum principle yields no information along the solution $\varphi^v(t)$ about the optimality of u(t). Such an arc of curve $\varphi^v(t)$ is called <u>"totally singular"</u>. The arc would be <u>singular</u> (but not totally singular) if there is an adjoint vector which is orthogonal to some (but not all) columns of B.

We can, therefore, state the following

<u>Theorem I</u> (Hermes). The Pfaffian form (VIII. 1. 4) is an exact differential if and only if φ^v is a totally singular arc.

On the contrary : φ^v <u>not</u> a totally singular arc implies the Pfaffian form (VIII. 1. 4) is <u>not</u> an exact differential, which implies that there exists $t_1 \geqslant t_o$ and u_φ such that $Z(t_1; \varphi^v, u_\varphi)$ is non-singular and the attainable set at time t_1 contains a neighborhood of the point $\varphi^v(t_1)$. The ="contrapositive" statement provides an interesting characterization of totally singular arcs: if for every $t_1 > t_o$ there exists points in every state space neighborhood of $\varphi^v(t_1)$ which are not attainable in time t_1 with L_2 controls, the arc φ^v is totally singular.

It may be noted that, strictly speaking, a totally singular arc satisfies the maximum principle. Therefore it may happen that such an arc is, indeed, time optimal.

A strong statement, referring to complete (not only local) controllability, is the following

Theorem II (Hermes). If the system (VIII. 1. 1) is not completely controllable at t_o then $Z(t; \varphi^v, u)$ is singular for all $t \geq t_o$, all possible u_φ and all reference trajectories φ^v, i.e. every solution φ^v is a totally singular arc.

Proof: any vector $h(t,x)$ orthogonal to the columns of $B(t,x)$ satisfies $h(t,x). B(t,x) \equiv 0$; hence, for any vector $v(t)$,

$$\frac{\partial}{\partial x}\left[h(t,x). B(t,x). v(t)\right] \equiv 0 ,$$

or

$$v(t). B^T(t,x). h_x(t,x) \equiv -h(t,x). B_x(t,x). v(t) .$$

Evaluation of this identity at $(t, \varphi^v(t))$, substitution into (VIII. 1. 5) and expansion of the left side yields

$$h_t + h. g_x + g. h_x^T \equiv v. B^T. (h_x - h_x^T) , \tag{VIII. 1. 6}$$

evaluated always at $(t, \varphi^v(t))$. This identity is necessary and sufficient condition for (VIII. 1. 4) to be an exact differential, i.e. φ^v a totally singular arc.

Now, if (VIII. 1. 1) is not completely controllable, the Pfaffian form $h(t,x)\,dx - h(t,x)\,g(t,x)\,dt$, $(h(t,x) . B(t,x) \equiv 0)$ is an exact differential, and

$$h_t(t,x) = - h(t,x)\ g_x(t,x) - g(t,x)\ h_x^T(t,x) ,$$

$$h_x(t,x) = - h_x^T(t,x) = 0 .$$

Evaluating these identities at $(t, \varphi^{v}(t))$ for an arbitrary control $v(t)$ shows that (VIII. 1. 6) is satisfied, hence every solution φ^{v} is a totally singular arc.

In connection with Pontryagin's maximum principle, the following intuitive considerations can also be made. The proof of the maximum principle is made, comparing, along a given solution $\varphi^{v}(t)$, the values of dx obtained with the (optimal) control $v(t)$, with those obtained by other choices of the control $u(t)$. These "possible perturbations" of the trajectory φ^{v} are then "transferred" along the solution φ^{v}, according to the " variation equation" $(\Delta x)' = f_x(t, \varphi^{v}(t), v(t)) \cdot \Delta x$, in order to obtain the corresponding variations of the final point of the trajectory. The possible final values of Δx form (in the linear approximation) a convex cone. The optimality assumptions implies that the final point is on the boundary of the attainable set ; therefore the convex cone of perturbations must lie on one side of a certain hyperplane (otherwise it would have the final point in its interior). This way the adjoint vector p is determined, such that at each point $(t, \varphi^{v}(t))$, the control is so chosen that $p. f(t, x, u) = \text{maximum}$. In the case of linear control $(f = g + B. u)$ and with a restrictive condition $u(t) \in U$ compact, this implies that in case $p. B \neq 0$, the optimal u has to belong to the boundary of U. But then, if we could choose values of u <u>not</u> restricted to U, we could take a <u>better</u> value for u (one which improve the optimality condition). This situation changes basically if, in the neighborhood of the final point, there would be points which were not attainable, even with dropping the restriction $u(t) \in U$. This is the case, as we have seen, for a totally singular arc, and then the maximum principle gives no information about optimality of the solution under consideration.

An interesting example of a singular problem (a problem which admits totally singular solutions) is given by Snow ([1]) in a recent paper. The following scalar equation is considered

$$x'' + a(t)\,x' + b(t)\,x + c(t) = u(t)\,,$$

with $|u| \leq 1$. The initial state is arbitrary, the desired final state is the origin and the optimality criterion is

$$J(u) = \int_0^T |u(t)|\,dt = \text{minimum.} \quad (x(T) = 0).$$

It is shown that this problem admits totally singular solutions if and only if $b(t) \equiv a'(t)$, so that the equation reduces to

$$\frac{d}{dt}\left[x' + a(t)\,x\right] = u(t) - c(t),$$

in which case the maximum principle yields no information. Nevertheless, the problem makes perfectly sense and has a solution. This solution is not unique, Snow goes further and determines, among all optimal solutions, the one which is also time-optimal.

A method which works for singular problems in the two-dimensional case, is the transformation of the curvilinear integral of the "cost" function in a double integral on a plane region. For this method we refer to the paper of Hermes and Haynes ([1]).

§ 2. Finite stability. Closely related to the concept of null-controllability, used in optimal control problems, is the finite stability. Let us recall that a point (t_o, x_o) is called null-controllable, if there exists an admissible control $u(t)$ such that the corresponding solution

$x(t)$ of the control equation , attains the origin in finite time.

The concept of finite stability was formulated by Roxin ([12]) for generalized dynamical systems, as seen in chapter III. Consider such a generalized dynamical system, determined by its attainability function $F(t, t_o, x_o)$, defined for $t, t_o \geqslant 0$, $x_o \in R^n$. Assume that the origin $x = 0$ is a <u>"strong point of rest"</u>, i.e. for any $0 \leqslant t_o \leqslant t$,

$$F(t, t_o, 0) = \{0\},$$

We then define the origin to be <u>finitely strongly stable</u>, if :

i) Given $t_o \geqslant 0$, $\varepsilon > 0$, there is $\delta = \delta(t_o, \varepsilon) > 0$ such that

$|x_o| \leqslant \delta$ implies $F(t, t_o, x_o) \subset S_\varepsilon$ for all $t \geqslant t_o$, where S_ε is the ε-neighborhood of the origin in R^n.

ii) For every $t_o \geqslant 0$ there is $\eta > 0$ such that if $|x_o| \leqslant \eta$, every motion $x(t)$ starting at (t_o, x_o), reaches the origin in finite time, i.e. $x(t) = 0$ for $t \geqslant T$, where T depends, in general, on (t_o, x_o) and on the particular motion $x(t)$.

We also define the origin to be <u>uniformly finitely strongly stable</u>, if :

i) Property (i) of the preceding definition holds with $\delta = \delta(\varepsilon)$ independent of t_o.

ii) There are $\eta > 0$ and $T > 0$, such that if $|x_o| \leqslant \eta$, $t_o \geqslant 0$, then for every motion $x(t)$ starting at (t_o, x_o), $x(t) = 0$ holds for all $t \geqslant t_o + T$.

It is quite obvious that properties (i) of both definitions are straightforward generalizations, for generalized dynamical systems, of the well known Liapunov stability and uniform stability of classical dynamical systems, through each point (t, x) there is only one motion $x(t)$ of the system , while for generalized dynamical systems, through each point (t, x) there are many motions, which determine the attainable set $F(., t, x)$. Here we demand that the classical stability condition holds for every motion through (t_o, x_o).

Properties (ii) in both definitions, generalize in some sense the asymptotic stability, but in a stronger version : solutions should not only approach asymptotically the origin, but reach it actually in finite time. This makes sense here, because there is no uniqueness of the solution through each point (in this case, through the origin).

We can also generalize the concept of stability , demanding that the classical property holds, not for all, but for some motion of the generalized dynamical system. Proceeding in that sense, we obtain the following definitions.

Assume that the origin is a "weak point of rest", i.e. $x(t) \equiv 0$ is a motion of the system.

The origin is said to be finitely weakly stable, if for every $t_o \geqslant 0$ there is $\eta(t_o) > 0$ such that $|x_o| \leqslant \eta$ implies the existence of a motion $x(t)$ which will be now denoted $x(t, t_o, x_o)$, such that:

i) $x(t_o, t_o, x_o) = x_o$

ii) $x(t, t_o, x_o) = 0$ for all $t \geqslant t_o + T$, where T depends, in general, on (t_o, x_o),

iii) $x(t, t_o, x_o) \to 0$ as $x_o \to 0$, for t_o fixed, uniformly in $t \in [t_o, +\infty)$.

Condition (iii) corresponds to the "weak Liapunov stability" in generalized dynamical systems.

Also:

The origin is said to be uniformly finitely weakly stable,

if in the definition of finite weak stability $\eta > 0$ and $T > 0$ can be taken independent of (t_o, x_o), and in (iii) $x(t, t_o, x_o) \to 0$ as $x_o \to 0$, uniformly for all $t_o \geq 0$, $t \geq t_o$.

It is apparent that the finite weak stability corresponds to the null-controllability of control systems. It is noteworthy that the so called "second method" of Liapunov for characterization of stability, can be applied in generalized dynamical systems for both cases of strong and weak type of stability. Roxin has shown this in recent papers ([4], [5], [6], [8], [9], [10], [11], [12]). Here we will give only, as an example and without proof, a theorem concerning finite weak stability.

<u>Theorem</u> (Roxin, [12]). Given an autonomous or periodic control system, a necessary and sufficient condition for the origin to be uniformly finitely weakly stable, with $T = T(\eta) \to 0$ as $\eta \to 0$, is the existence of a scalar valued function $v(t, x)$ with the following properties :

i) $v(t, x)$ is defined for $x \in R^n$, $t \geq 0$, where it is lower semicontinuous ;

ii) $v(t, 0) = 0$;

iii) there is a strictly increasing continuous function $v_1(r)$, defined for $r \geq 0$, $v_1(0) = 0$, such that $v(t, x) \geq v_1(|x|)$;

iv) $v(t, x) \to 0$ as $x \to 0$, uniformly in $t \geq 0$;

v) there are constants $\lambda > 0$, $c > 0$ such that at all points (t, x) where $0 < v(t, x) \leq \lambda$, the condition $D_+ v(t, x) \leq -c < 0$ holds.

Here

$$D_+ v(t,x) = \liminf_{\tau \to t+} \left\{ \frac{v(\tau, y) - v(t,x)}{\tau - t} \; ; \; y \in F(\tau, t, x) \right\}$$

is a kind of lower right derivative of $v(t,x)$ for the control system determined by $F(t, t_o, x_o)$.

Note that in this theorem we demand more than the uniform finite weak stability, demanding the condition $T(\eta) \to 0$ for $\eta \to 0$. Systems which have this property are called, in the paper mentioned, "nicely controllable".

Markus ([1]) has given sufficient conditions for the whole R^n to be "null - controllable" or, as we would say here, for "weak finite stability in the large".

References.

Antosiewicz, H.A. :

[1] Linear control systems, Archive Rat. Mech. Anal. 12(1963), 313-324.

Balakrishnan, A.V. :

[1] Optimal control problems in Banach spaces, J. SIAM Control 3 (1965), 152-180.

Barbashin, E.A. :

[1] On the theory of generalized dynamical systems, Uch. Zap. M.G.U. 135 (1949), 110-133.

Bellman, R., Glicksberg, I. and Gross, O. :

[1] On the bang-bang control problem, Quarterly Appl. Math. 14 (1966), 11-18.

Blackwell, D. :

[1] The range of certain vector integrals, Proc. Amer. Math. Soc., 2 (1951), 390-395.

Bliss, G.A. :

[1] Lectures on the calculus of variations, The Univ. Chicago Press, 1946.

Bonnesen, T. and Fenchel, W.:

[1] Theorie der Konvexen Körper, Springer, Berlin, 1934 and Chelsea Publ. Co., N. York, 1948.

Bouligand, G. :

[1] Introduction à la géomètrie infinitésimale directe, Paris, 1932.

Bushaw, D. :

[1] Optimal discontinuous forcing terms, Contrib. Theory of Nonlinear Oscillations, IV. Princeton Univ. Press, 1958.

[2] Dynamical polysystems and optimization, Contrib. to differential Equations, 2 (1963) , 351-365.

Carathéodory, C. :

[1] Variationsrechung, 5th chapter of the book of R. v. Mises, Die Differentialgleichungen und Integralgleichungen der Mechanik und Physik, Vieweg & Sohn, Braunschweig, 1930. This chapter is also reproduced in Carathéodory's "Gesammelte Werke".

Castaing, C. :

[1] Sur une extension du théorème de Liapunov, Comptes rendus Acad. Sci. Paris, t. 260 (1965) , 3838-3841.

Cesari , L :

[1] An existence theorem in problems of optimal control , J.. SIAM Control, 3 (1965) , 7-22.

[2] Existence theorems for optimal solutions in Lagrange and Pontryagin problems, J. SIAM Control 3(1965), 475-498.

Coddington, E. A. and Levinson, N. :

[1] Theory of ordinary differential equations, McGraw Hill, N. York, 1955.

Conti, R. :

[1] Contributions to linear control theory, J. Differential Equations, 1 (1965) , 427-445.

Fattorini, H. O . :

[1] Time optimal control of solutions of operational differential equations, J. SIAM Control 2 (1964) , 54-69 .

[2] Control in finite time of differential equations in Banach space, Comm. Pure Appl. Math. XIX (1965) , 17-34.

Filippov , A.F. :

[1] Differential equations with multi-valued discontinuous right hand side, Mat. Sborn . (1960) .

[2] On certain questions in the theory of optimal control, Vestnik Moskov. Univ. Ser. 1 Mat. Meh. 2 (1959) , 25-32 (russian) ; english translation in J. SIAM Control 1 (1962), 76-84.

Fukuhara, M. :

[1] Sur le systèmes des équations différentielles ordinaires, Proc. Imperial Acad. Japan 4 (1928) , 448-449 .

[2] Sur l'énsemble des curves intégrales d'un système d'équations différentielles ordinaires, Proc. Imperial Acad. Japan 6 (1930) , 360-362 .

Gambill , R.A. :

[1] Ceneralized curves and the existence of optimal controls, J. SIAM Control 1 (1963) , 246-260 .

Gamkrelidze, R.V. :

[1] On sliding optimal regimes, Dokl. Akad. Nauk SSSR , 143 (1962), 1243-1245 (russian) ; english translation in Soviet Math. Dokl. 3 (1962) , 390-395 .

[2] On some extremal problems in the theory of differential equations with applications to the theory of optimal control, J. SIAM Control 3 (1965) , 106-128 .

Gottschalk, W.H. and Hedlund, W.A. :

[1] Topological dynamics, Amer. Math. Soc. Colloqu. Publ. 36, Providence , 1955.

Halkin , H. :

[1] On the necessary conditior for optimal control of nonlinear systems, J. Anal. Math. XII (1964) , 1-82 .

[2] Topological aspects of optimal control of dynamical polysystems, Contributions to Differential Equations 3 (1964) ,

[3] A generalization of LaSalle's bang-bang" principle , J. SIAM Control 2 (1964) , 151-159 .

[4] On generalization of a theorem of Liapunov, Journ. Math. Anal. Appl.

[5] Some further generalization of a theorem of Liapunov , Archive Rat. Mech. Anal, 17 (1964) , 272-277.

Halmos P.R. :

[1] The range of a vector measure, Bull. Amer . Math. Soc. 54 (1948) , 416-421.

Hartman, P. :

[1] Ordinary differential equations, Wiley, N. York, 1965.

Hermes, H. :

[1] Controllability and the singular problem, J. SIAM Control, 2 (1964), 241-260.

[2] The equivalence and approximation of optimal control problems Brown Univ. Tech. Rep. 65.2, Center Dynam. Syst., Providence, 1965.

Hermes, H. and Haynes, G. :

[1] On the nonlinear control problem with oontrol problem, J. SIAM Control 1 (1963) , 85-108.

Kalman, R.E. :

[1] Contributions to the theory of control, Bol. Soc. Mat. Mexicana 5 (1960) , 102-119.

[2] On the general theory of control systems, Proc. First Int. Congress IFAC, Moscow, 1960, Butterworth, London , 1961, Vol. 1, 481-492 .

[3] Canonical structure of lin—ear dynamical systems, Proc. Nat Acad. Sci USA, 48 (1962), 569-600.

[4] Mathematical description of linear dynamical systems, J. SIAM Control 1 (1963) , 152-192.

Kalman, R. E. Ho, Y.C. and Narendra, K. S. :

[1] Controllability of linear dynamical systems, Contributions Differential Equations 1 (1963) , 189-213.

Kamke, E. :

[1] Zur Theorie der Systeme gewöhnlicher Differentialgleichungen II, Acta Math. 58 (1932) , 57-85.

Kneser, H . :

[1] Über die Lösungen eines Systems gewöhnlicher Differentialgleichungen das der Lipschitz'schen Bedingung nicht genügt, S. B. Preuss . Akad. Wiss. Phys. Math. Kl. (1923), 171-174.

LaSalle, J. P. :

[1] Time optimal control systems, Proc. Nat. Acad. Sc. USA 45 (1959) , 573-577.

[2] The time optimal control problem, Contrib. Theory Nonlinear Oscillations V (1960) , Princeton Univ. Press. , 1-24.

Leitmann, G. :

[1] Some geometrical aspects of optimal processes , J. SIAM Control 3 (1965) , 53-65 .

Levinson , N. :

[1] Minimax, Liapounov and bang-bang, J. Differential Equations 2 (1966) , 74-101.

Liapunov, A.A. :

[1] Sur les fonctions-vecteurs complétement additives, Izv. Akad. Nauk SSSR , Ser. Mat. 8 (1940), 465-478 (russian) ;

McShane, E.J. :

[1] Generalized curves, Duke Math. J. 6 (1940) , 513-536.

[2] Necessary conditions in generalized curve problems in the calculus of variations, Duke Math. J. 7 (194) , 1-27 .

Marchaud, A. :

[1] Sur les champs continus de demi-cônes convexes et leurs intégrales Comptes rendus Acad. Sci. Paris 199 (134), 1278-128

[2] Sur les champs de demi-cônes et les équations différentielles du premier ordre, Bull. Soc. Math. France 62 (1934), 1-38.

[3] Sur les champs continus de demi-cônes convexes et leurs intégrales, Compositio Math. 3 (1936) , 89-127.

Markus , L. :

[1] Controllability of nonlinear processes, J. SIAM Control 3 (1965) 78-90.

Markus, L. and Lee, E.B. :

[1] Optimal control for nonlinear processes, Archive Rat. Mech. Anal. 8 (1961) , 36-58.

Neustadt, L.A. :

[1] The existence of optimal controls in the absence of convexity conditions, J. Math. Anal. Appl. 7 (1963) , 110-117.

[2] Optimization, a moment problem and nonlinear programming, J. SIAM Control 2 (1964) , 33-53.

[3] A general theory of minimum-fuel space trajectories, J. SIAM Control 3 (1965) , 317-356.

Olech, C. :

[1] A note concerning extremal points of a convex set, Bull. Acad. Polon. Sci., Ser. Math. Astron. Phys. XIII (1965), 347-351.

[2] A note concerning set-valued measurable functions, Bull. Acad. Polon. Sci., Ser. Math. Astron. Phys. XIII (1965), 317-321.

[3] Extremal solutions of a control system, J. Differential Equations, 2(1966), 74-101.

Pliś, A. :

[1] Trajectories and quasitrajectories of an orientor field, Bull. Acad. Polon. Sci., Ser. Math. Astron. Phys, XI (1963), 369-370

Pontryagin, L. S., Gramkrelidze, R. V. And Boltyanskii, V. G. :

[1] The theory of optimal processes - The maximum principle, Izv. Akad. Nauk SSSR, Ser. Mat. 24 (1960), 3-42.

Pontryagin, L. S. Gamkrelidze, R. V., Boltyanskii, V. G. and Mischchenko, E. F. :

[1] Mathematical theory of optimal processes, (orig. russian), english transl. edit. Interscience, N. York. 1962.

Pugh, C. :

[1] Cross sections of solution funnels, Bull. Amer. Math. Soc. 70 (1964), 580-583.

Rishel, R. W. :

[1] An extended Pontryagin principle for control systems whose control laws contain measures, J. SIAM Control 3 (1965), 191-205.

Roxin, E. :

[1] Reachable zones in autonomous differential systems, Bol. Soc. Mat. Mexicana 5 (1960), 125-135.

[2] A geometric interpretation of Pontryagin's maximum principle, RIAS Tech. Rep. 61.15 (1961) and Proc. Int. Conf. Nonlinear Oscillations, Colorado Springs, 1961 .

[3] The existence of optimal controls, Mich. Math. J. 9 (1962) , 109-119.

[4] Axiomatic theory of control systems, RIAS Tech. Rep. 62-12, (1962) .

[5] Axiomatic foundation of the theory of control systems, 2nd. Int. Congress of IFAC, Basel, 1963.

[6] Stability in general control systems, J. Differential Equations 1 (1965), 115-150.

[7] On generalized dynamical systems defined by contingent equations, J. Differential Equations 1 (1965) , 188-205.

[8] Stabilität in allgemeinen Regelungssystemen, 3rd. Conf. Nonlinear Oscillations, Berlin, 1964.

[9] Local definition of generalized control systems, Mich. Math. J. (1966) .

[10] On stability in control systems, J. SIAM Control 3 (1965) , 357-372.

[11] On asymptotic stability in control systems, to appear in the Rendic. Circolo Mat. Palermo.

[12] On finite stability in control systems, to appear in the Rendic. Circolo Mat. Palermo.

Roxin, E. and Spinadel V. :

[1] Reachable zones in autonomous differential systems, Contrib. Differential Equations 1 (1963), 275-315.

Sedzıwy, S:
1 On the trajectories and quasitrajectories of a two-dimensional orientor field, Bull. Acad. Sci. Polon., Ser. Mat. Astron. Phys. XII (1964), 157-159.
[1] On the trajectories and quasitrajectories of a two-dimensional orientor field, Bull. Acad. Polon. sci., Ser. Mat. Astron. Phys. XII (1964) , 157159.

Schmaedecke, W. W. :
[1] Optimal control theory for nonlinear vector differential equations containing measures, J. SIAM Control 3 (1965), 231-280.

Snow, D. :
[1] Singular optimal controls for a class of minimum effort problems, J. SIAM Control 2 (1964) , 203-219.

Turowicz, A. :
[1] Sur les trajectoires et quasitrajectoires de systèmes de commande nonlinéaires, Bull. Acad. Polon. Sci., Ser. Math. Astron. Phys. X (1962) , 529-531.
[2] Sur les zones d'émission des trajectoires et quasitrajectoires des systèmes de commande nonlinéaires, Bull. Acad. Polon. Sci., Ser. Math. Astron. Phys. XI (1963), 47-50.
[3] Remarque sur la définition des quasitrajectoires d'un système de commande nonlinéaire, Bull. Acad. Polon. Sci., Ser. Math. Atron. Phys. XI (1963) , 367-368.

Warga, J. :
[1] Relaxed variational problems, J. Math. Anal. Appl. 4 (1962), 111-128.
[2] Necessary conditions for minimum in relaxed variational problems, J. Math. Anal. Appl. 4 (1962) , 129-145.
[3] Minimax problems and unilateral curves in the calculus of

Wazewski, T. :

[1] Système de commande et équations au contingent, Bull. Acad. Poln. Sci., Ser. Math. Astron. Phys. IX (1961), 151-155.

[2] Sur une condition d'existence des fonctions implicites measurables, Bull. Acad. Poln. Sci., Ser. Math. Astron. Phys. IX (1961), 861-863.

[3] Sur une condition équivalente à l'équation au contingent, Bull. Acad. Polon. Sci., Ser. Math. Astron. Phys. IX (1961), 865-867.

[4] Sur la sémicontinuité inférieuredu "tendeur" d'un ensemble compact variant d'une facon continue, Bull. Acad. Polon. Sci. Ser. Math. Astron. Phys. IX (1961), 869-872.

[5] Sur une généralization de la notion des solutions d'une équation au contingent, Bull. Acad. Polon. Sci. Ser. Math. Astron. Phys. X (1962), 11-15.

[6] Sur les systèmes de commande nonlinéaire dont le contredomaine de commande n'est pas forcément convexe, Bull. Acad. Polon. Sci., Ser. /Math. Astron. Phys. X (1962), 17-21.

[7] Sur quelques définitions équivalentes des quasitrajectoires des systèmes de commande, Bull. Acad. Polon. Sci., Ser. Math. Astron. Phys. X (1962), 469-474.

[8] On certain conditions of existence of periodic trajectories and quasitrajectories of the control system of differential equations. Non linear vibration problems 5 (1963) 376-377.

[9] Sur un système de commande dont les trajectoires coincident avec les quasitrajectoires du systeme de commande donné Bull. Acad. Polon. Sci., Ser. Math. Astron. Phys. XI (1963), 101-104.

Young, L. C. :

[1] Necessary conditions in the calculus of variations, Acta Math. 69 (1938), 246.

Zaremba, S. C. :

[1] Sur une extension de la notion d'équation différentielle, Comptes rendus Acad. Sci. Paris 199 (1934), 545-548.

[2] Sur les équations au paratingent, Bull. Sci. Math. 60 (1936), 139-160.

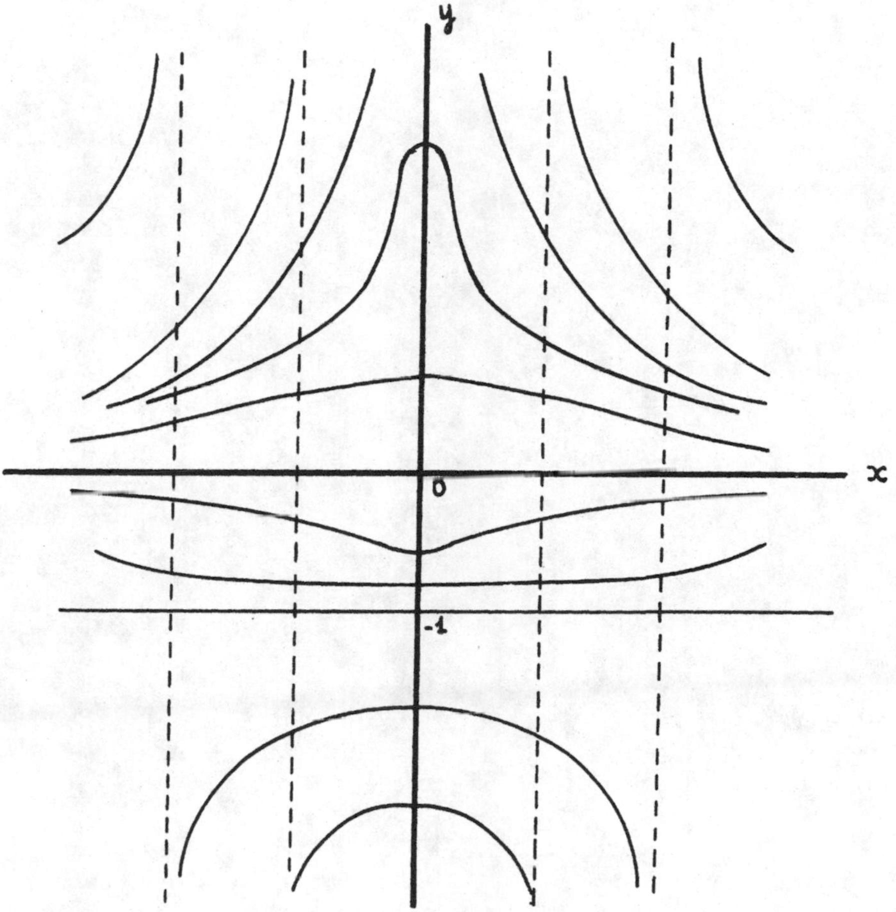

Fig. II.1.

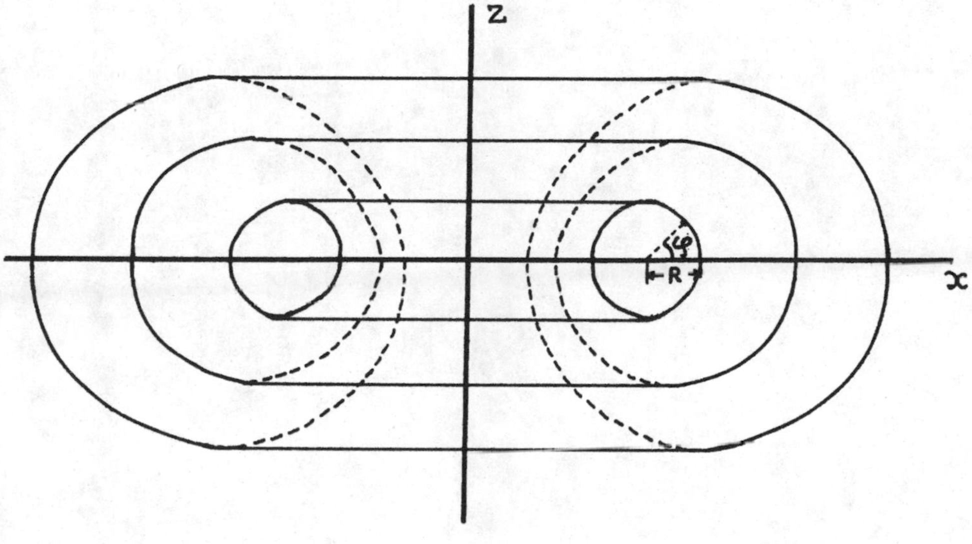

Fig. II.2.

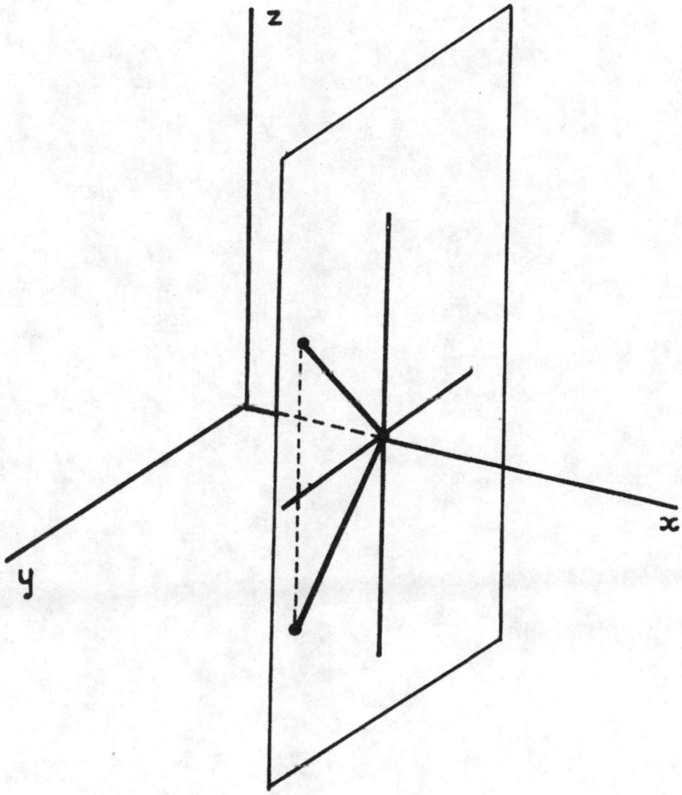

Fig. II.3.

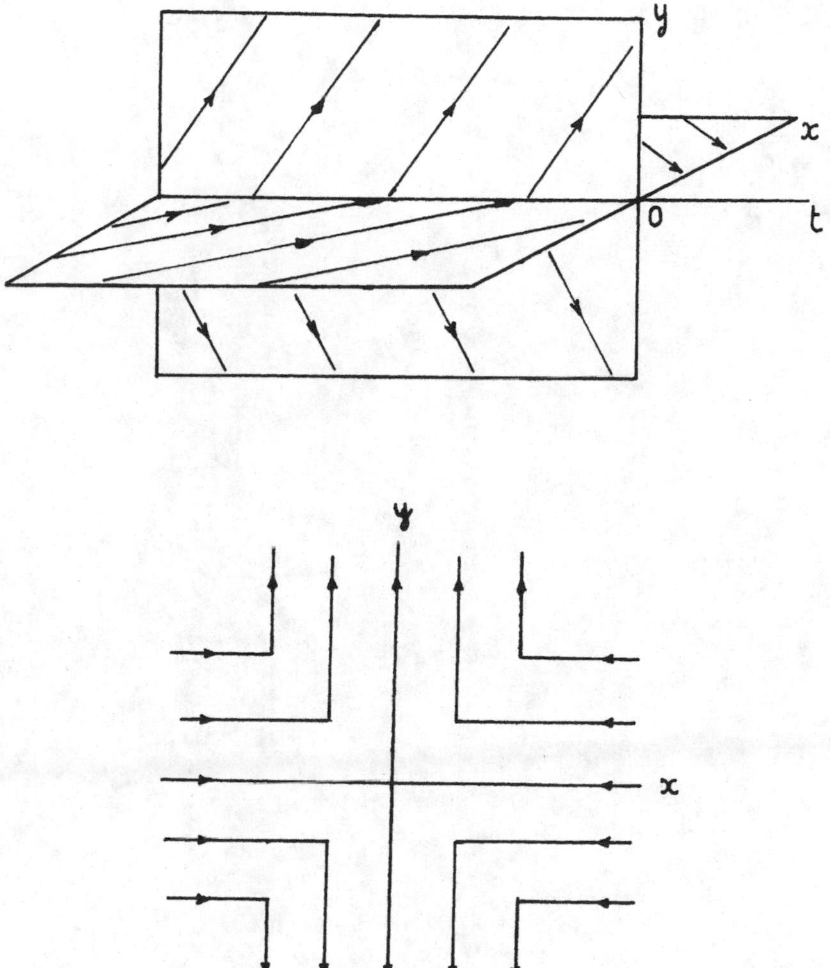

Fig. III.1.

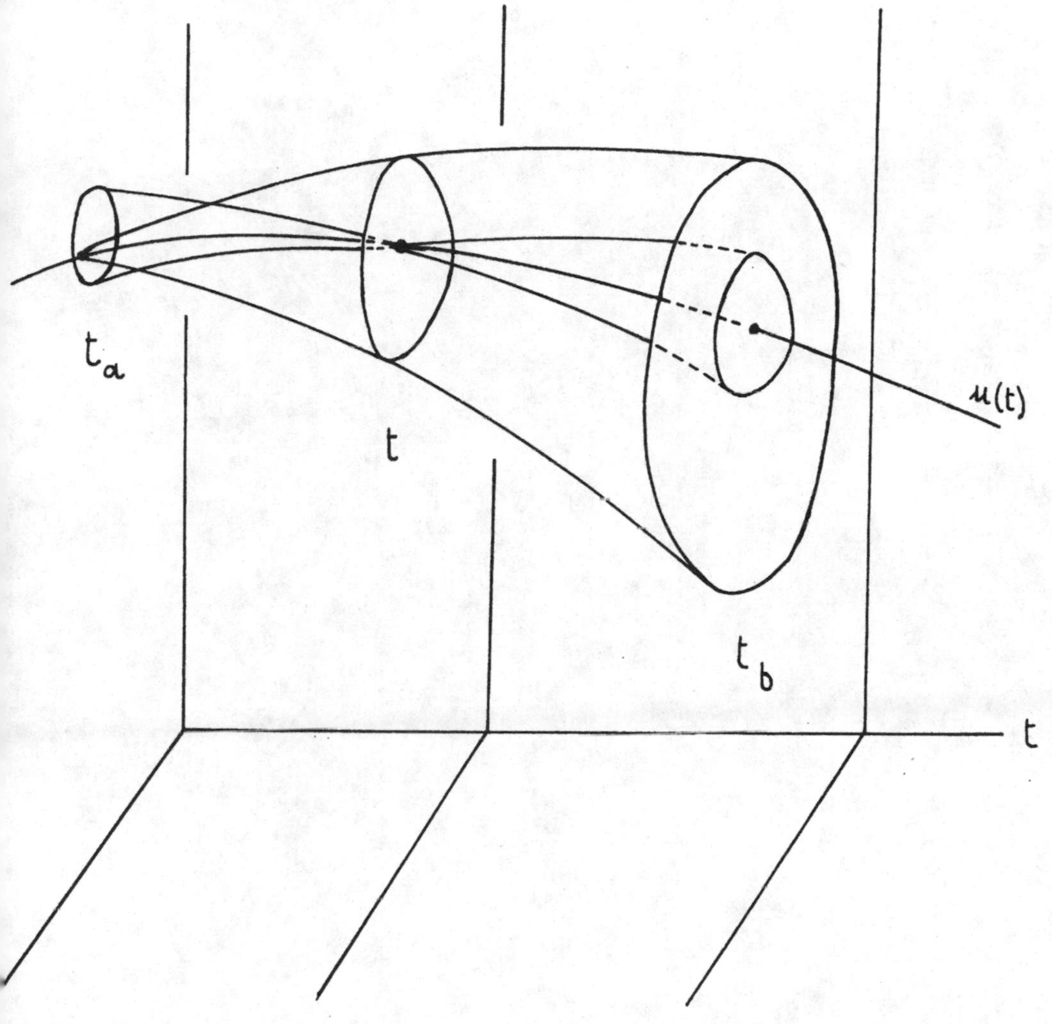

Fig. III.2.

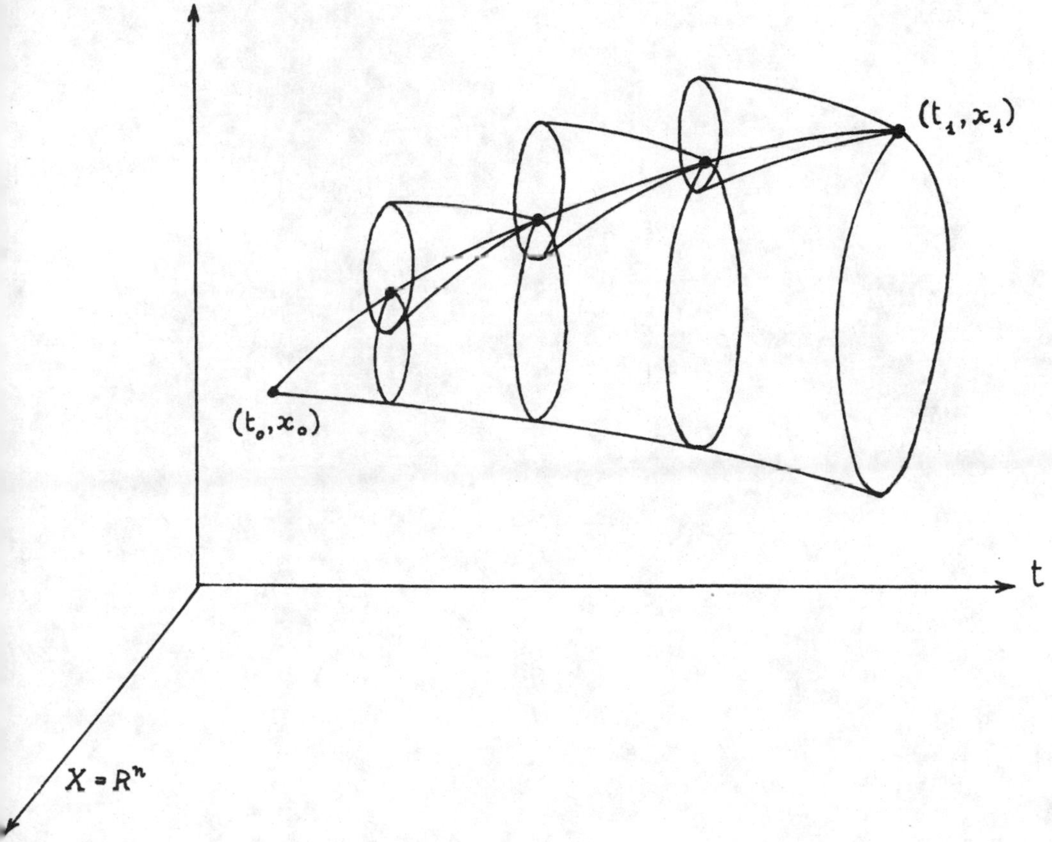

Fig. V.1.

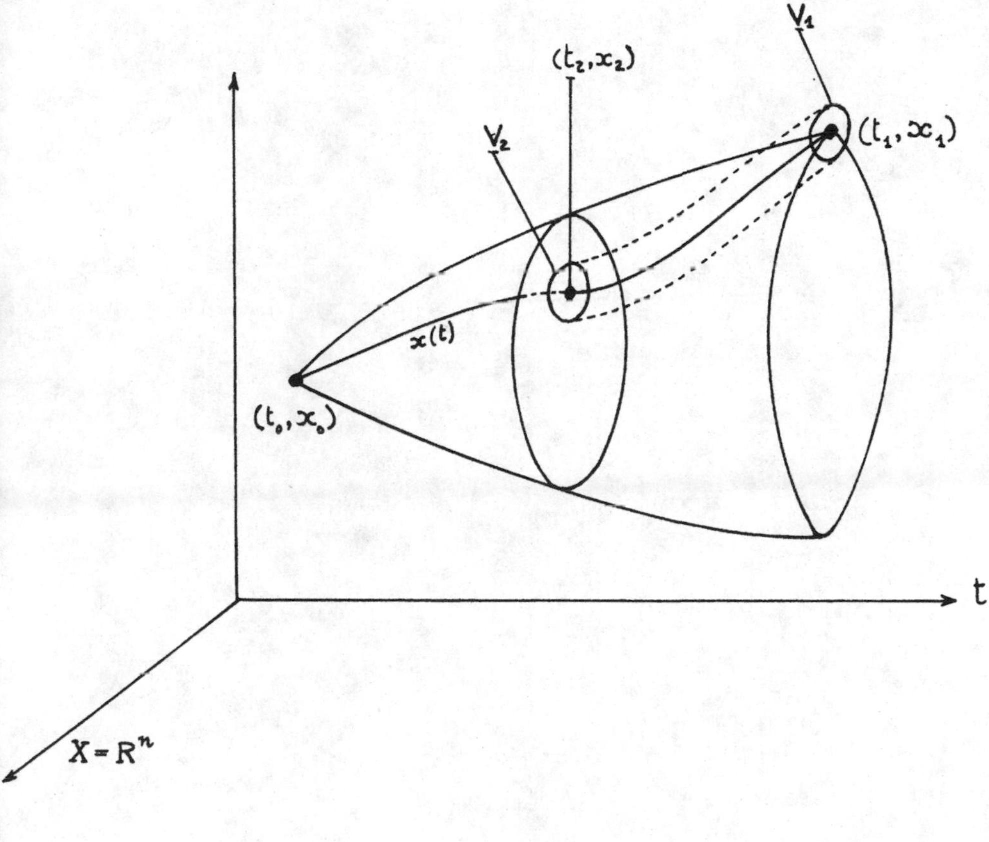

Fig. V.2.

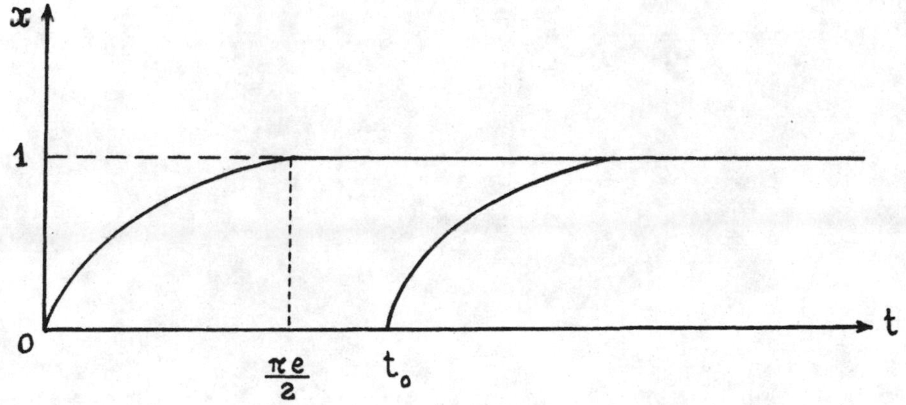

Fig. V.3.

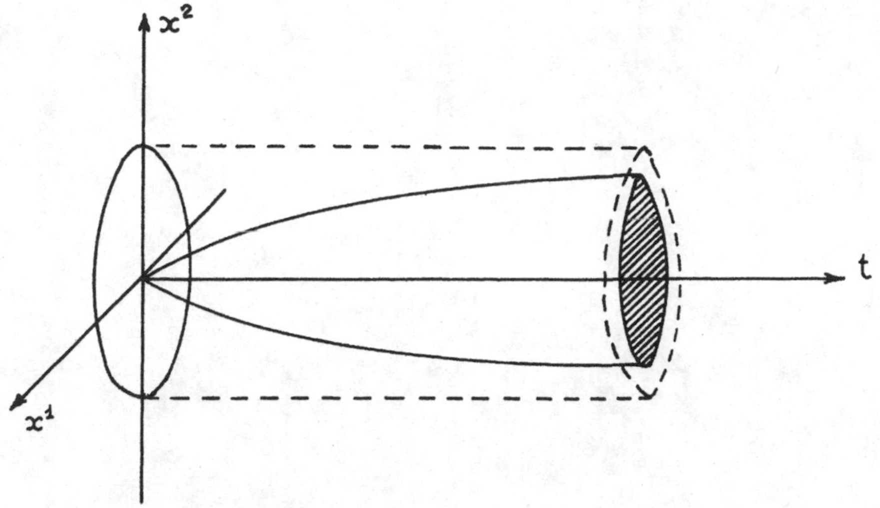

Fig. VI.1.

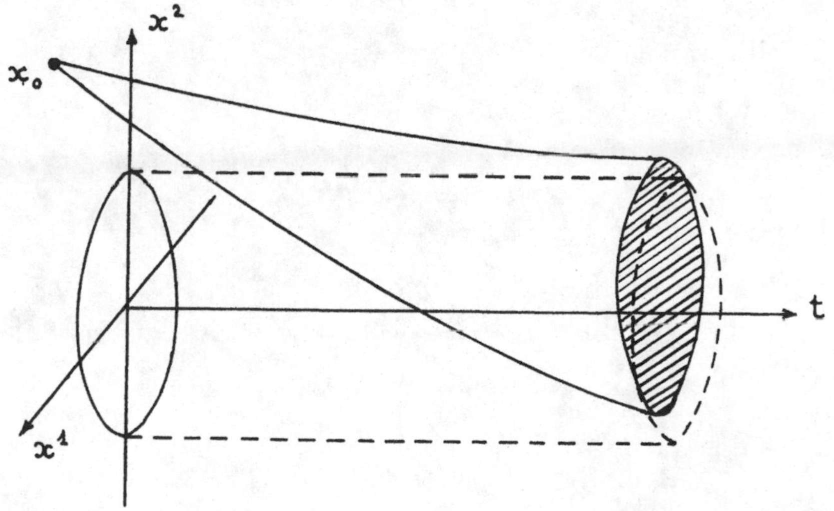

Fig. VI.2.

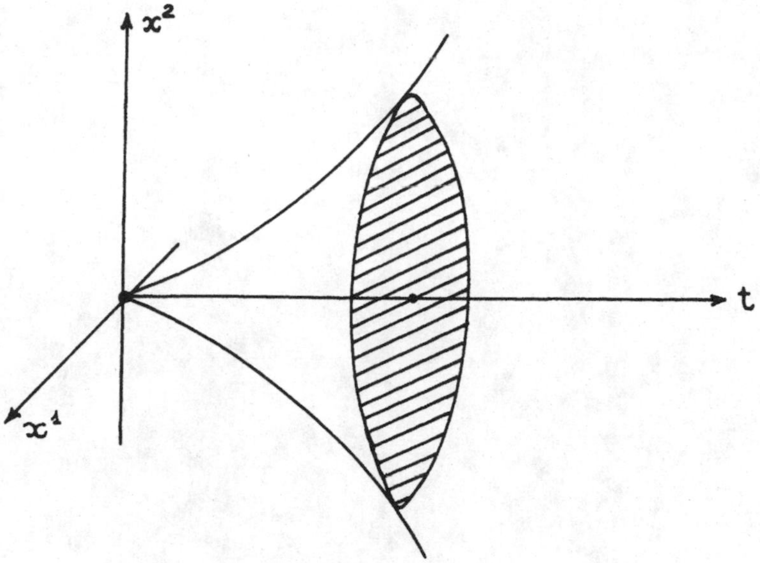

Fig.VI.3.

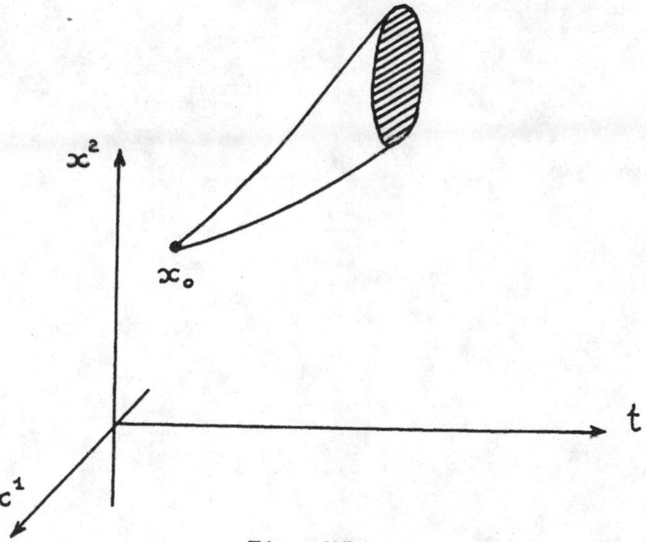

Fig. VI.4

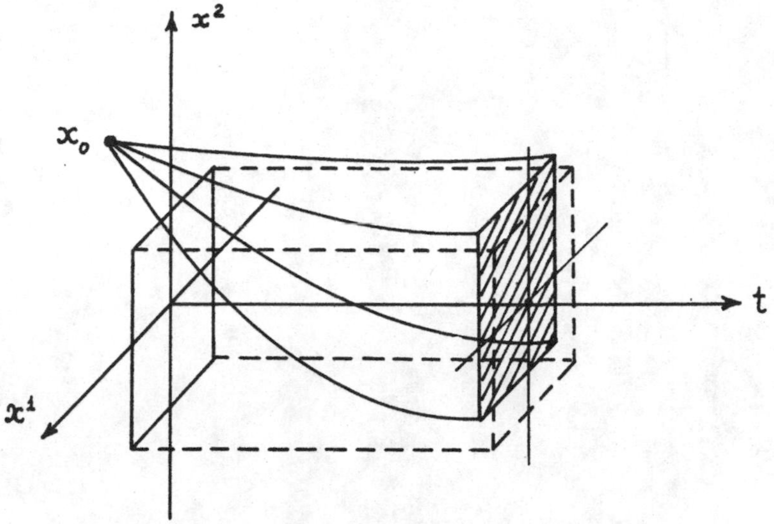

Fig. VI.5.

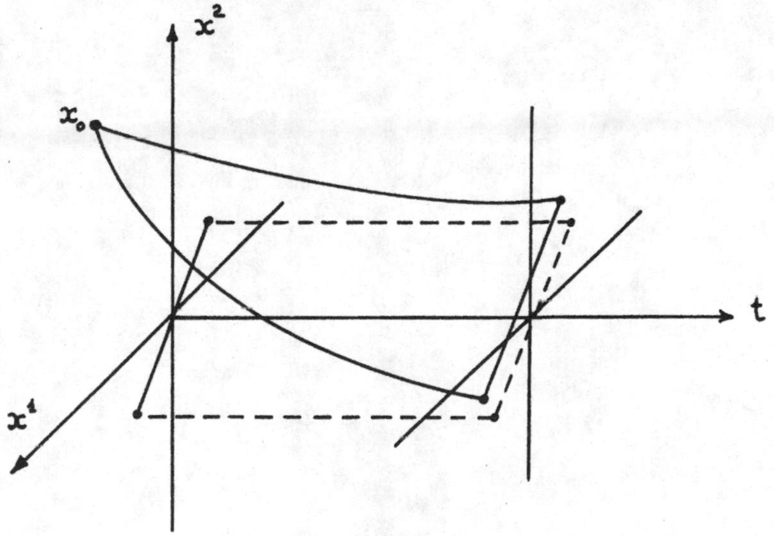

Fig. VI.6.

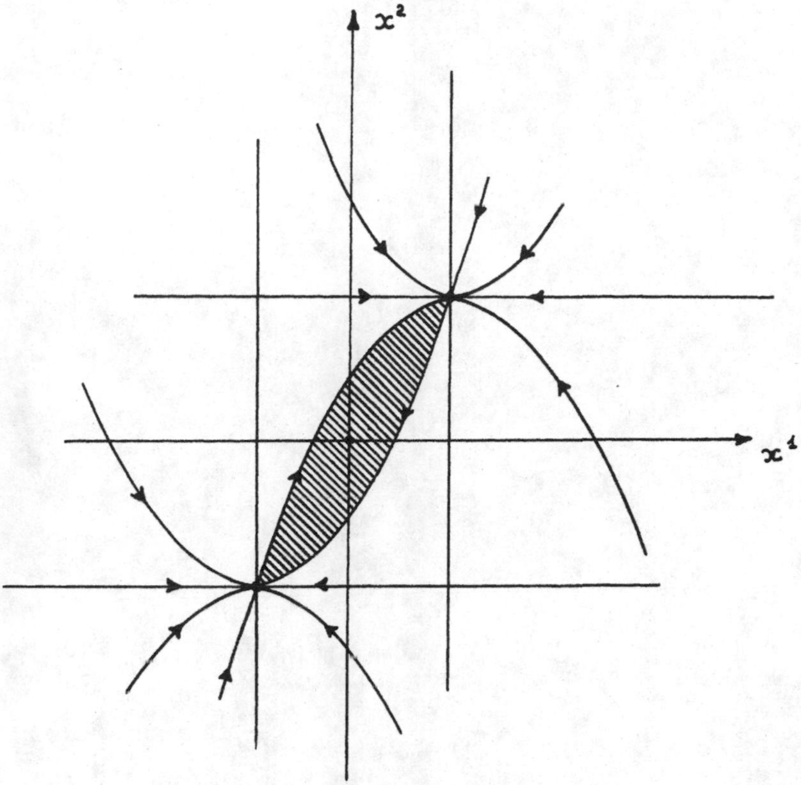

Fig. VI.7.

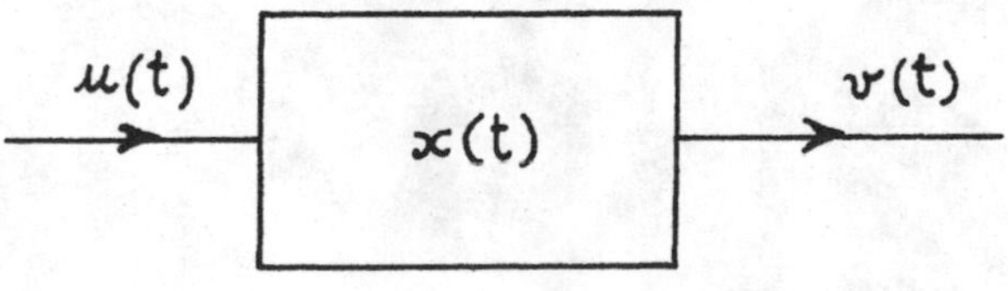

Fig. VI.8.

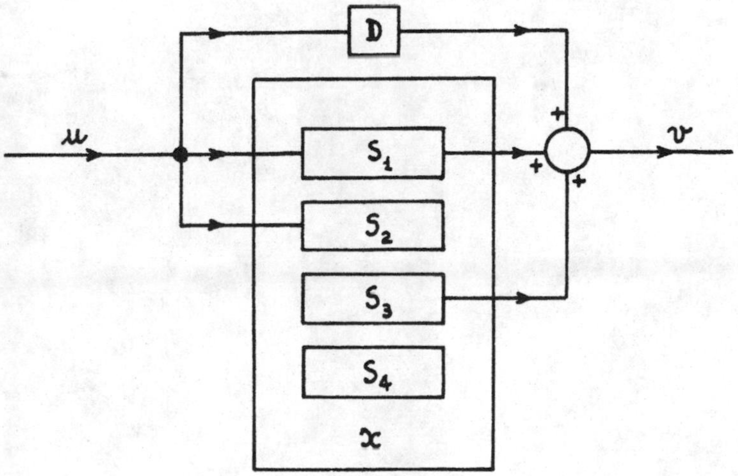

Fig. VI.9.

Printed in the United States
By Bookmasters